CIRIA C604

London, 2004

CDM Regulations –

work sector guidance for designers

Second edition

Taking account of the ACoP HSG224 (2001)

Prepared by Ove Arup and Partners

Updated by Alan Gilbertson

CIRIA *sharing knowledge ■ building best practice*

Classic House, 174–180 Old Street, London EC1V 9BP
TELEPHONE 020 7549 3300 FAX 020 7253 0523
EMAIL enquiries@ciria.org
WEBSITE www.ciria.org

Summary

The Construction (Design and Management) Regulations 1994 (CDM) apply to nearly all construction work undertaken in the UK and for all projects the designers will have duties under CDM. This guide helps any person or organisation acting as a designer to meet their obligations in an effective manner. Each of the legal duties is explained in detail and advice given on how they may be effectively discharged. This new edition takes account of the ACoP publication HSG224 (2001). In addition to giving advice on a designer's legal duties, the guide explains how the work may be carried out in an effective, proportionate manner and guidance is given on the difficult issues which arise in practice. The issue of "who is a designer for the purposes of CDM" is addressed.

The guide starts by explaining to designers the need for action and what is required of them by CDM; this includes details of the different types of accident and health issues involved in construction. The major part of the content consists of the examination of health and safety issues arising in the 39 different work sectors considered. For each section, hazards are identified and discussed, examples of risk mitigation are given and references provided to related issues in other sections and to other publications that may assist. Finally, guidance is given about documenting design decisions and sources of further information of a general nature.

CDM Regulations – work sector guidance for designers

1st edition author: Ove Arup and Partners
2nd edition reviser: Gilbertson, A

CIRIA publication C604 1st edn © CIRIA; 1997 2nd edn revisions © CIRIA 2004
First published 1997 as Report 166; updated and revised 2004 ISBN 0-86017-604-5

Keywords		
health and safety, construction management, Construction (Design and Management) Regulations, CDM, construction work, demolition, design, designers, hazards, maintenance, risk		
Reader Interest	**Classification**	
Construction industry designers, architects, civil engineers, structural engineers, services engineers, surveyors, consultants, local authorities	AVAILABILITY	Unrestricted
	CONTENT	Original research
	STATUS	Committee guided
	USER	Construction sector designers, (engineers, architects, surveyors, project managers) and planning supervisors, construction industry professionals

Published by CIRIA, Classic House, 174-180 Old Street, London EC1V 9BP.

About this guide

THE CDM REGULATIONS

The Construction (Design and Management) Regulations 1994 (known as the CDM Regulations or the Regulations in this guide) affect nearly all construction work of any significance and were a major development in health and safety legislation. The Regulations place duties upon all designers and this guide is designed to assist in fulfilling those duties.

The CDM Regulations build upon earlier health and safety legislation by imposing a framework of duties so that all the parties to a construction project must consider health and safety. The Regulations have an Approved Code of Practice (ACoP) which has legal status and must be referred to alongside the CDM Regulations.

THE ACoP

Advice in the document *Managing Health and Safety in Construction: Construction (Design and Management) Regulations 1994*, HSG224, which contains the Code of Practice and related guidance, has particular significance as follows.

- The Code of Practice itself is in bold type; it gives practical advice on how to comply with the law with respect to the matters covered, although you may use other methods if they enable you to comply with the law.
- Other guidance is in normal type; much of it is an interpretation of good practice (using terms such as *should* and *need to*) but some is an interpretation of the law (*must*).

In practice, it is safer to apply all of HSG224. As HSG224 is commonly referred to as the ACoP and is numbered by paragraph, regardless of whether the text is approved code or guidance, reference to the paragraphs in HSG224 is to "ACoP para N".

Note that the current ACoP (HSG224) replaced the original 1994 ACoP (L54) in 2001. It is substantially different and this was the prime driver for the second edition of this publication.

READERSHIP

Work sector guidance for designers is written for those wishing to develop a full understanding of the designer function under CDM. It focuses upon the management of hazard and risk but also provides information about designers' CDM duties in general and reference to other CIRIA guidance on related CDM issues, which designers should be aware of. The term *designer* is a defined term under the Regulations and has a broad meaning, going beyond the traditional meaning to include anyone who makes input to design decisions. This is discussed in the guidance. It should also be noted that during the stages of a project, different designers will be active; initially consultants may be making conceptual decisions and working up schemes, but later trade contractors will be designing their work and the main contractor will be designing site systems including lifting arrangements.

Each designer will have his own area of decision-making, which will affect risks on site. Designers need to concentrate on the decisions that they can influence while being aware of the concerns of other designers, which may be affected by their actions.

A general understanding of the Regulations and the ACoP is assumed. For further information see the section "Sources of further information". In particular, designers may refer to two companion CIRIA documents:

- C602 – *Practical guidance for clients and clients' agents,* which provides information about when CDM applies and the roles of parties under CDM
- C603 – *Practical guidance for planning supervisors,* which provides detailed information about the role of the planning supervisor and his interaction with designers.

KEY POINTS

It must be understood that anyone who acts as a designer (including a client who imposes specific requirements) has duties as a designer under CDM.

All those who contribute to design decisions must consider the hazards and risks involved. This requires an understanding of construction work and the types of accidents and health issues that are involved.

This report provides a wealth of accessible information about health and safety issues in a wide range of construction activity. The consideration of these matters must be an integral part of the design decision process. Each designer has a part to play, working as a member of the design team who must make decisions related to health and safety in a co-ordinated manner.

The object of CDM is to embed health and safety management into projects. All designers have a role to play and must communicate with others to provide information about health and safety for contractors to take into account, as they plan and execute work on site.

A designer's duties under CDM require considerable common sense and openness in order to relate to the other dutyholders in a constructive manner. Duties must be carried out in a manner that is proportional to the type of project and the likely level of risk. Information prepared by designers for the purposes of CDM must:

- focus on health and safety information that competent people would not reasonably anticipate
- be specific to the project
- reflect the level of risk and complexity
- be concise.

Particular care is needed to prepare concise, focussed documentation that is relevant to the project in question. In this way it will be of immediate use to the people who need to use it. To achieve this, the use of lengthy standardised or off-the-shelf catch-all documents should be avoided. The aim is to produce relevant documents that are proportionate to the project and its risks, and hence are useful and cost-effective.

HOW TO USE THIS GUIDE

The guide has been structured so that it can be read from cover to cover or consulted by those who simply need to dip in for specific information. It is divided into sections that focus on work sectors and for each project only the relevant sectors need be consulted. The guidance is designed to provoke thought and to improve understanding and knowledge; it cannot provide complete information for every circumstance. As with all checklist guidance, the user must ask "is there anything else, for this particular project?"

Chapter 1 – Introduction

This section introduces the duties of a designer under CDM, defines the key terms and suggests how the guidance may be used in managing hazard and risk. It provides information about over-arching issues, the planning and sequencing of construction, common causes of accidents and the major health hazards in construction.

Chapter 2 – Work sector guidance

A wide range of construction activities are examined, by work sector. For each sector, major hazards are identified and discussed, risk mitigation measures are proposed and reference to further guidance is provided.

Chapter 3 – Documenting design decisions on individual projects

Advice is given about the need for decisions to be documented, including the statutory duty to provide health and safety information to others.

Chapter 4 – Glossary

Chapter 5 – Sources of further information

Acknowledgements

The research work leading to this publication was carried out by Ove Arup and Partners, under contract to CIRIA. The project supervisor was Allan Delves assisted by Paul Craddock and publication design was by Steven Groak. CIRIA's research manager for the project was David Churcher.

The book was updated in 2003 to align with the amendments to the Regulations in 2000 and the revised ACoP published in 2001. The contractor for this work was Alan Gilbertson of Gilbertson Consultants Limited and CIRIA's research manager was David Storey.

First edition (1996–97) – published as R166

Work sector authors

The work sector authors were all drawn from the staff of Ove Arup & Partners:

David Anderson	Howard Porter
Barry Austin	Gordon Puzey
Bob Cather	Mark Rudrum
Paul Craddock	Peter Ross
Allan Delves	Eric Wilde
Roger Drayton	Mel Wheeler
Roger Howkins	Library staff

Sketches and graphics

The sketches were prepared by Fred English to whom a special vote of thanks is recorded. The graphics were undertaken by Denis Kirtley, Trevor Slydel and Martin Hall.

Project Steering Group

The research was guided by a Steering Group, established by CIRIA to advise on technical scope and presentation of content. This comprised:

Mr A D Rogerson (Chairman)	McDowells
Mr S Bell	Royal Institute of British Architects, CONIAC
Professor D Bishop	Construction Industry Council
Mr E Criswick	DoE Construction Sponsorship Directorate
Mr P Gray	Scott Wilson Kirkpatrick & Co Ltd
Mr H Hosker	Building Design Partnership
Mr D Lamont/Mr N Thorpe	Health and Safety Executive
Dr M Lockwood	Construction Industry Council
Mr B Mansell	Institution of Civil Engineers, CONIAC
Mr I Neil	Acer Consultants (latterly Hyder Consulting Ltd)
Mr R Oughton	F C Foreman & Partners
Mr J Read	WS Atkins Consultants Ltd
Mr P White/Mr T Hetherington	Health and Safety Executive
Mr D Williams	Balfour Beatty/Senior Safety Advisers Group

In addition, detailed technical comment on individual groups of work sectors was obtained from the following Review Groups, established by CIRIA:

Group A

Mr J Ahern	J Ahern & Associates
Mr D Brattle	Lovell Construction
Mr K Dew	John Pelling & Partners
Mr E Kulikowski	John Laing Construction
Mr B Mansell	Institution of Civil Engineers
Mr P White	Health and Safety Executive

Group B

Mr S Corbet	G Maunsell & Partners
Mr P Gray	Scott Wilson Kirkpatrick & Co Ltd
Mr J Hatto	Bovis Construction Ltd
Mr R Herring	Kier Engineering Services
Mr B Norton, CBE	Federation of Civil Engineering Contractors
Mr N Thorpe	Health and Safety Executive

Group C

Mr D Brown	Steel Construction Institute
Mr D Chapman	Bovis Construction Ltd
Mr K Dew	John Pelling & Partners
Mr B Norton CBE	Federation of Civil Engineering Contractors
Mr J Read	WS Atkins Consultants Ltd
Mr M C Shurlock	Kier Engineering Services Ltd
Mr B Todd	Kier Engineering Services Ltd
Mr P White	Health and Safety Executive
Mr P Williams	British Constructional Steelwork Association

Group D

Mr W Adams	Tilbury Douglas Construction
Mr S Bell	RIBA, CONIAC
Mr K Dew	John Pelling & Partners
Mr J Elkington	NBS Services
Mr H Hosker	Building Design Partnership
Mr C Lawrence	Tilbury Douglas Construction
Mr R Nanayakkara	BSRIA
Mr R Oughton	F C Foreman & Partners
Mr S Thompson	The Boyd Partnership, RIAU
Mr N Thorpe	Health and Safety Executive
Mr D Watson	WSP Group

Group E

Mr D Cowan	Laing Civil Engineering
Mr J Modro	Balfour Beatty Civil Engineering
Mr J Read	WS Atkins Consultants Ltd
Dr L Smail	Union Railways (latterly Bechtel Ltd)
Mr N Thorpe	Health and Safety Executive
Mr P York	Highways Agency

The CI/SfB classification codes in each work sector were checked by Mr J Cann of NBS Services and Annette O'Brien of Ove Arup & Partners.

CIRIA and Ove Arup & Partners would like to express their thanks to all those individuals and organisations who took part in the consultation for this project, including the Steering Group and Review Groups.

Funding

The research project was funded by DoE Construction Sponsorship Directorate, the Health and Safety Executive, Institution of Civil Engineers Research & Development Enabling Fund, CIRIA's Core Programme, and from a contribution-in-kind from the research contractor.

Second edition (2003) – this publication

Project Steering Group

The updating was guided by a Steering Group, established by CIRIA to advise on technical issues. CIRIA and Gilbertson Consultants Limited would like to express their thanks and appreciation to all members of the Project Steering Group and their organisations. This comprised:

Dr O Jenkins (Chairman)	CIRIA
Mrs E Bennett	Habilis
Mr G Briffa	HSM Ltd (for the Association of Planning Supervisors)
Mr P Craddock and team	Arup
Mr D Lambert	Kier Group plc
Mr T Lyons	Taylor Woodrow
Mr R Whiteley	WS Atkins
Mr S Wright	Health and Safety Executive

Additional assistance was given with review of the work sector guidance texts by the following:

Miss R Davies	Black and Veatch
Mr S Ayub	Buro Happold
Mr J Worthington	Balfour Beatty
Mr R Cather	Arup
Mr P Gray	Scott Wilson
Mr I Catherwood	HR Wallingford
Mr R Smith	Shepherd Construction
Mr R McClelland	Alfred McAlpine

Contents

LIST OF FIGURES

LIST OF TABLES

1 Introduction

1.1 THE PURPOSE, SCOPE AND LIMITATIONS OF THIS GUIDANCE

This guidance has been produced to assist designers to comply with Regulation 13(2) and 13(3) of the Construction (Design and Management) Regulations 1994, (CDM). This Regulation requires designers to ensure that any design prepared for the purposes of "construction work" during construction, maintenance, repair or demolition includes adequate regard to the avoidance and reduction of hazards, protection of all affected and protection of individual workers. This guidance touches on duties other than those under Regulation 13 (2) and 13 (3), to set them in context. The scope of a designer's responsibility under the CDM Regulations is examined fully in the Approved Code of Practice (ACoP), which all practicing designers should be familiar with; see in particular paras 103–137.

The purpose of this guidance is to help designers to manage health and safety risks arising from their designs, while meeting their statutory duty, in the following ways.

1. It provides information on the major hazards associated with each of the specific work areas.

2. It gives key considerations, prompts and examples to encourage designers to think about the peculiarities of their own specific projects and to identify and prioritise project-related hazards, with respect to the stages of design.

3. Once the health and safety hazards have been identified, it provides signposts to the types of responses that designers can make to alter their design so as to avoid, reduce or control the hazards.

4. It gives cross-references to other sectors of construction work addressed by the report, which may be related in terms of health and safety, and to external texts from which more detailed information may be obtained.

The guidance has been drafted for designers who are active in design practice in any construction discipline and need rapid access to the key considerations connected with a particular construction work sector. Within each work sector, what the designer can achieve is influenced by the design stage and previous decisions that have been made, either by the designer or by others.

The scope of this report is restricted to 39 work sectors, chosen following industry consultation and intended to represent those areas of construction work considered to contain the most significant and common hazards and risks to the health and safety of construction workers and others (eg members of the public and building occupiers). However, these work sectors by no means address all hazards in construction. Every designer has to think through the design options to identify all the hazards that can reasonably be foreseen and then act accordingly to avoid or reduce them.

A number of the hazards mentioned specifically in the work sectors, or those that designers may further identify, may be outside a particular designer's scope of control or may be issues that a principal contractor would deal with. However, they are included so that designers may be as fully informed as possible about the range of potential hazards. Those not familiar with site practices might find *Health and safety in construction*, HS(G)150, a useful introduction to the main health and safety hazards in construction work.

The designer's duties relate to considering hazards that can reasonably be identified and to provide information so that contractors and others can be made aware of them. Designers are not required to outline methods of dealing with hazards but should explain any particular construction assumptions that have been made as part of the design.

Readers need to appreciate that the guidance in this document is not intended to repeat information available elsewhere, but to include reference to it, so it is not intended that this handbook can be read entirely in isolation.

1.2 DEFINITIONS (SEE ALSO GLOSSARY)

1.2.1 Designer and construction work

The CDM Regulations define the terms *designer* and *construction work* with much wider meanings than those common in construction usage. See Section 1.4.3 for an explanation of the term designer and see Section 2 of CIRIA Publication C602 for further explanations on construction work.

1.2.2 Stages of a project

The terms *concept stage*, *scheme stage* and *detail stage* have been used to subdivide the design process. It should be noted that much design is done after the appointment of the principal contractor.

The designer's response to any hazards identified will vary according to the stage of design development. There is more flexibility to avoid hazards at the start of the design process during concept stage, than during the detail stage when control measures may be more appropriate for dealing with any remaining hazards.

1.3 THE NEED FOR ACTION TO IMPROVE HEALTH AND SAFETY IN CONSTRUCTION

The chances of being killed during a working lifetime on construction sites are 1 in 300 and the chances of being disabled by serious illness or injury is much greater than for workers in other industries. In addition, members of the general public are also injured and killed by construction work. Figures from the Health and Safety Executive indicate that on average (2000/1 to 2002/3) 85 workers are killed on sites each year. An analysis of this figure by types of accident is given in Table 1.1. Although the total number and mix of fatalities varies each year, the nature of accidents has remained broadly the same.

Table 1.1 *Indicative breakdown of fatalities according to type of accident*

Percentage of total	Type of accident
47%	Falls from a height
15%	Struck by a falling/flying object, eg during machine lifting of materials
10%	Contact with electricity or an electrical discharge
7%	Struck by a moving vehicle
7%	Trapped by something collapsing or overturning
3%	Contact with moving machinery or material being machined
11%	Other causes

There is also a large incidence of health problems amongst construction workers and Table 1.2 gives estimates for numbers suffering ill-health as a result of certain hazards. These numbers dwarf the figures for injuries and on-the-job fatalities and in many cases, the effects may not be felt until years of exposure has taken place.

Table 1.2 *Indicative annual estimates of numbers suffering from work-related ill health in the construction industry*

Hazard (possible resulting disease or condition)	Annual estimates Lower limit	Upper limit
Asbestos (mesothelioma, asbestosis, lung cancer)	200 (deaths)	250 (deaths)
Musculoskeletal injury (back disorders, work-related upper limb disorders, lower limb disorders)	30 000	48 000
Respiratory disease (bronchitis, emphysema, asthma, pneumoconiosis, sinusitis, influenza)	3500	23 000
Skin disease	3100	10 500
Noise (deafness and ear conditions)	1000	5800
Vibration conditions	200	400

From the figures in these tables, it is clear that great improvements can be made. To do so, contributions need to be made by clients, designers and contractors. In particular designers need access to guidance that will help them identify the hazards arising from their design and then to avoid, reduce or control these hazards through the design process.

1.4 THE DESIGNER'S DUTIES UNDER CDM

1.4.1 Background

The Health and Safety at Work etc Act 1974 puts duties on employers, employees and the self-employed. Section 3, for example, states that:

> *It shall be the duty of every employer to conduct his undertaking in such a way as to ensure, so far as is reasonably practicable, that persons not in his employment who may be affected thereby are not exposed thereby to risks to their health or safety.*

The Management of Health and Safety at Work Regulations 1999 take this further and specifically require an assessment of risk and appropriate control measures as a key feature of effective health and safety management. The Construction (Design and Management) Regulations 1994 extend these principles to designers of structures.

The Management of Health and Safety at Work Regulations 1999 place broad health and safety duties upon all employers, self-employed people and employees. They overlap with CDM and in some areas extend the requirements of CDM, as outlined in Appendix 2 of the ACoP.

1.4.2 The importance of design in managing health and safety in construction

The Regulations place the designer at the centre of health and safety, jointly with the contractor, because the work to be done is of the designer's choosing.

> The ACoP states (paras 103–106) "Designers are in a unique position to reduce the risks that arise during construction work, and have a key role to play in CDM... Designers' earliest decisions can fundamentally affect health and safety... [Designers] need to consider the health and safety of those who will maintain, clean and eventually demolish a structure... designers have to weigh many factors [as well as health and safety]... including cost, fitness for purpose, aesthetics, buildability and environmental impact.

All designers therefore need to be familiar with the Regulations and the ACoP, or they will not be able to undertake the legally enforceable duties placed upon them by CDM.

1.4.3 Who are designers?

It is important to appreciate who is classed as a designer. As well as the design team appointed by a client (or a design and build contractor), the ACoP (paras 109–111) makes it clear that in CDM the term *designer* has a broad meaning, including:

- those who analyse, calculate, do preparatory design work, design, draw, specify, prepare bills of quantities
- those who arrange for their employees (or others under their control) to do design
- anyone who specifies or alters a design or who specifies the use of a particular method of work or material.

Nearly all participants in a project could be designers, whether it be a client specifying a building layout or a type of construction, or a subcontractor deciding how to work.

The guidance provided in this report is for **ALL** designers. Note that designers are required to work as a team, assisted by the planning supervisor. Decisions that are not in the control of one designer, due to the contractual framework, will be in the control of another designer. Each project will be different and it is therefore not appropriate for the Regulations to set out exactly how the various designers on a project will interact. As the design progresses, designers will need to take account of decisions already made and information provided by others. They need to raise questions where necessary and make further decisions using their own expertise. Examples of areas of design that require careful co-ordination include:

- all interfaces between elements, particularly where there is a degree of interaction
- interfaces between permanent works and temporary works
- works that will be left in a temporary condition while other work continues.

1.4.4 The role of lead designer

In addition to individual designers, there is a need for a lead designer, responsible for promoting co-ordination and co-operation throughout the life of a project.

1.4.5 Understanding the context

Designers need to understand the context within which they are performing their CDM duties; the ACoP and CIRIA publications C602 and C603 may be referred to. The following table lists helpful sections.

Table 1.3 *Sources of guidance on particular topics*

Topic	ACoP Guidance	CIRIA Guidance
Scope of CDM and the roles of the dutyholders	Paras 21–51	C602 Chapter 4
Role of the client, particularly with respect to providing information	Paras 13–16, 52–102	C602 Chapter 4
Role of the planning supervisor, particularly with respect to co-ordination of designers	Paras 138–155	C603 Chapter 6
Competence and resources	Paras 191–213	C603 Chapter 5
Health and safety plan	Paras 229–247 and appendix 3	C603 Chapters 7 and 8
Health and safety file	Paras 248–267 and appendix 4	C603 Chapter 9
Different types of procurement	Paras 65–72	C602 Chapter 9

1.4.6 Designers' duties

Designers' duties under CDM are contained in Regulation 13 and amplified in the ACoP Paras 103–137. **All designers should read this guidance and become familiar with it**. HSE Construction Information Sheet No. 41 also gives guidance on designers' duties.

Designers have to:

- ensure that their clients are aware of their duties under CDM
- seek to eliminate or reduce hazards and risks in their design
- co-operate with the planning supervisor and other designers
- provide information for the pre-tender health and safety plan and health and safety file.

The duty to advise clients of their duties under CDM requires that designers are well informed about CDM in general. CIRIA publication C602 provides information relevant to the client's role. Designers should note that clients are required to provide the planning supervisor/designers with information that they could reasonably be expected to obtain, which will be relevant to the management of hazards and risks. This requirement may assist in persuading a reluctant client to commission surveys (in conjunction with an explanation of the benefits such work will bring to the project).

The duty to co-operate with the planning supervisor is explored in CIRIA publication C603, which provides CDM guidance focused upon the role of the planning supervisor.

1.4.7 Designers' management of hazards and risks

The Regulations require (Regulation 13(2)) that:

"Every designer shall:

(a) ensure that any design he prepares and which he is aware will be used for the purposes of construction work includes among the design consideration adequate regard to the need -

(i) to ***avoid foreseeable risks*** *to the health and safety of any person at work carrying out construction work or cleaning work in or on the structure at any time, or of any person who may be affected by the work of such a person at work;*

(ii) to ***combat at source*** *risks to the health and safety of any person at work carrying out construction work or cleaning work in or on the structure at any time or of any person who may be affected by the work of such a person at work, and*

(iii) to ***give priority to measures*** *which will protect all persons at work who may carry out construction work or cleaning work at any time and all persons who may be affected by the work of such persons at work over measures which only protect each person carrying out such work."*

Regulation 13(3) qualifies this as requiring "the design to include only matters referred to therein to the extent that it is reasonable to expect the designer to address them at the time the design is prepared and to the extent that it is otherwise practicable to do so."

Limitations to the designers' duties are specifically covered in the ACoP paras 131 and 136–137:

- It is not necessary to point out every hazard (but designers must point out significant hazards, ie those that are not likely to be obvious, are unusual or are likely to be difficult to manage).
- Only foreseeable hazards and risks need be dealt with.
- Construction methods do not need to be specified unless the design assumes or requires a particular sequence of work or a competent contractor might need the information (but designers must always understand how their designs can be built safely).
- There is no direct responsibility for contractors' performance on site (although unsafe practices should always be pointed out if they are witnessed).
- There is no specific legal requirement to keep records of the process of design risk assessment.

A structured approach to the principles of prevention and protection will often point to sensible answers. Professional judgement is required, as designers cannot remove all risks. However, just because a particular hazard (eg falls from an open edge) is common throughout the industry, this does not mean that designers should dismiss the possibility that they can act to influence it. In many cases, work that is done routinely by contractors can be affected by designers. For example, contractors regularly work on fragile roofs and workers regularly fall through. By choosing more robust materials, this hazard is avoided.

Designers need to consider all risks and where a project includes unusual risks, designers need to provide information and the planning supervisor needs to pass on this information in the pre-tender health and safety plan. Where specific action, on either an unusual or a common risk, has been taken and the benefit might be lost if this were not brought to the attention of the contractor, this information should also be included. There is no need to provide information about operations that the designer has been unable to avoid and that normally are reliably performed by a competent contractor. The significant project-specific risks should be flagged up.

Industry attention has been mainly focused on the impact of the CDM Regulations on new construction and renovation. Repair-and-maintenance accounts for a substantial proportion of the industry's turnover and for more than its proportional share of reported accidents. By including access and a safe place of work for maintenance as a design consideration, many maintenance hazards could be significantly reduced. Major hazardous construction tasks are readily appreciated, but many examples stem from too little thought having been given to routine maintenance tasks – cleaning work in atria, possible only by erecting a scaffold; restricted access in plant rooms best served by contortionists; fragile roofs with no access provision, and so on. Clearly much can, and should, be done by designers. Further guidance is given in CIRIA publication *Safe access for maintenance and repair* (C611). Risk analysis and the principles of prevention and protection will point the way.

1.4.8 Design risk assessment (DRA)

Designers need to examine ways in which hazards can be avoided or mitigated or, if neither is possible, designed so that the level of risk is acceptable (given proper controls), applying the principles of prevention and protection. This process is known as design risk assessment (DRA) and follows the hierarchy as set out in Regulation 4 of the Management of Health and Safety at Work Regulations 1999, reproduced in Appendix 2 of the ACoP.

Design risk assessment is an iterative process whereby a designer can logically identify, assess and manage risks. Shown in paras 121–123 of the ACoP as follows:

- **identify the hazards** in a proposed design
- **eliminate each hazard**, if feasible (or substitute a lower risk hazard)
- **reduce the risk** during construction work – this includes cleaning, maintenance or demolition
- **provide information** necessary to identify and manage the remaining risks.

This involves making judgements between possible courses of action and may significantly influence design. The approach should be structured to suit the particular project, including consideration of foreseeable hazards, the site and the local environment.

There are several methods for risk assessment, both qualitative and quantitative. To some degree all are subjective and arbitrary but may be useful – provided the method used is appropriate for the purpose and its limitations are recognised. Precise estimates of risk are not normally possible, as they are impractical and time consuming, and a lack of data nearly always makes quantitative analysis impossible.

Although risk assessment will usually be qualitative and based upon a balanced view, there may be occasions when a more advanced numerical form of analysis is appropriate, such as HAZAN (hazard analysis). This form of analysis is normally focused upon controlling hazardous situations or combinations of unusual circumstances in process plant. Where designers are familiar with their use and the circumstances are right, such techniques may be applied to CDM assessment work, but this will only occasionally be appropriate.

If the work to be done is sufficiently similar to other jobs that have been assessed for risks, detailed re-assessment may not be required. It may be appropriate to modify the original assessment in light of the new circumstances. However, when relying on previous assessments it is essential to ensure that all assumptions about the work and the circumstances in which it will be undertaken are relevant to the situation under consideration. This is mentioned in the ACoP para 124.

Guidance about risk assessment can be found in a wide variety of documents. Much of this is orientated towards risk assessment under the Management of Health and Safety at Work Regulations. Designers need to anticipate risks that may arise when constructing a design or when maintaining or even demolishing it many years later. Normally risk analysis of workplaces is based on what is already happening there. In construction the opposite is usually the case – tasks are not repetitive or routine and workplaces are not purpose-designed. Designers have to assess a likely scenario by using their experience, imagination and professional judgement. This must be informed by an appreciation of the hazards and risks to be avoided, mitigated or accepted and controlled.

Designers should not fall into the trap of leaving risk assessment until the end of the design, ie simply identifying the residual risks. Brain-storming at the start of a job will frequently identify the major issues that require further consideration as the design progresses. Hazard checklists may be useful as an aid, but are no substitute for experience and judgement. Each project is unique and has to be thought about in the light of the circumstances.

1.4.9 Identification of hazards

The first and essential step is to identify the **hazards**. There are two categories to consider:

1. hazards that are likely to be within the recognition of a competent contractor, but which might be reduced by designers' initiative
2. hazards that are likely to be outside the recognition of a competent contractor.

It is important to:

- identify and eliminate hazards from an early stage
- identify inter-disciplinary issues where two or more design elements interface
- consider all the detailed activities entailed in constructing the design and the risks that might arise from their interaction.

The ACoP paras 127–129 provides a checklist of examples as pointers on hazard recognition. See also Chapter 2 and Appendix A in this report and Section 1.7.4 below.

1.4.10 Assessing risk

The likelihood of an accident arising from a particular operation on a particular site during a particular afternoon is low. It would be illegal for a contractor to expose operatives to probable accidents, or for designers to design so that there is a significant risk. Moreover, operations such as finishing, plastering and glazing, which might be perceived as relatively safe, account for a large number of reported injuries and fatalities. The best course for designers is to consider what practical decisions they may take to make accidents less likely, in their judgement, than to seek for guidance from imperfect data.

1.4.11 Balancing the consequences

Making design decisions entails achieving an appropriate balance between:

1. the probable outcome of a risk materialising, whether
   - immediate (a death), evident in the short term (disabling) or in the medium to long term (health damage)
   - affecting a few or many individuals
   - creating very different types of damage, eg death, injury, damage to health

   and
2. the cost and inconvenience arising from sacrificing other design objectives, for example:
   - clients' functional requirements
   - cost (initial and whole-life)
   - programme
   - aesthetics
   - functionability
   - durability
   - environment.

Design need not be dominated by the requirement to avoid all risks at all cost, but consideration of risk must be a significant influence.

The measures that contractors can take on their own behalf to protect their workers (eg temporary edge protection, PPE) should be discounted by designers at this stage. Even though contractors will control risks by the application of well-known precautions, designers should always consider how hazards can be avoided or risks reduced. The designer's goal should be to reduce the total risk to be managed by contractors and to make the residual risks easier to manage, where this is possible. As the ACoP demonstates by example in paras 127–129, there are many areas of design where designers can make a difference.

1.4.12 Designers' contribution to risk control

Risk control measures should preferably be collective rather than personal. In descending order of effectiveness, the hierarchy of risk control involves:

- changes that eliminate a hazard
- substitution of a less hazardous design feature
- enclosure – isolation, barriers, guarding or segregation, all of which are designed to separate people from the hazard
- reduced exposure – changes that reduce the time individuals are exposed to a risk, or the number of people exposed
- safe systems of work, together with suitable training and supervision
- written procedures, and the provision of information, instruction, warnings, signs and/or labels
- use of personal protective equipment (PPE).

Failure by designers to address a hazard may mean that contractors have to adopt less effective measures such as the use of PPE. The designers' contribution to risk control therefore lies primarily in the scenario they set for the contractors. Designers select the risks that contractors must manage and the provision of information about residual risk (although necessary, see below) is of importance only once the key design decisions that affect risk have been made.

1.4.13 Provision of information

Designers have always provided information for construction, and under CDM they also need to provide information for the health and safety file, to guide and inform those responsible for future work (cleaning, maintaining, repairing, modifying, adapting and demolishing) on the structure.

The ACoP (para 130) clarifies the information that designers need to supply:

- hazards that remain in the design and the resultant risks
- any assumptions about working methods or precautions so that people carrying out the construction work can take them into account.

The ACoP (para 133) notes that the information must be "clear, precise and in a form suitable for the users", for example by:

- notes on drawings (the best solution in most cases where the information is not long or complicated)
- supporting documents if necessary, referenced from the notes on the drawings
- a register or list of significant hazards with suggested control measures
- suggested construction sequences showing how the design could be erected safely, where this is not obvious.

Limitations to and exclusions from a designer's duties under CDM, with respect to the provision of information, are specifically covered in the ACoP para 131 and 136–137.

<u>Para 131</u>: Designers do not need to mention every hazard or assumption (when providing information to others) as this can obscure the significant issues, but they do need to point out significant hazards. These are not necessarily those that result in the greatest risks, but those that are:

- unlikely to be obvious to a competent contractor or other (competent) designers
- unusual, or
- likely to be difficult to manage effectively.

Unless there are aspects directly influenced by the design solution, designers would not normally need to provide information about everyday hazards that could not be eliminated, for example those connected with:

- materials in common use such as cement and sand
- working at height
- open excavations
- site transport movements
- the use of scaffolding
- standard information included in regulations or codes of practice.

<u>Para 132</u>: Designers always need to provide information for significant hazards, such as those listed under (c) in Appendix B

Para 136: Designers are not required to :

- take into account or provide information about unforeseeable hazards and risks
- specify construction methods, except where the design assumes or requires a particular construction or erection sequence or where a competent contractor might need information.

Para 137: Designers are not legally required to keep records of the process they use to reach a safe design.

All designers should understand these requirements and their implications, as explained in the ACoP.

Information for construction will be provided on drawings and other documents and inserted into the health and safety plan. Information for the future will be inserted into the health and safety file, and shown on other documentation (such as drawings), which is to be held in the file or as supporting documentation.

The ACoP provides guidance about the contents of health and safety plans in paras 229–247 and Appendix 3, and of health and safety files in paras 248–267 and Appendix 4.

Guidance on the documentation of design decisions on individual projects is provided in Chapter 3 of this guide.

1.4.14 Audit trail of decisions

A planning supervisor should encourage the design team to keep records of crucial health and safety decisions. This is not a requirement of the Regulations, but should it be necessary to revise a design or to backtrack, as often happens during the later stages of design development, a record would avoid reconsidering health and safety matters long since decided. Records may also demonstrate how CDM duties were carried out, should there be an investigation.

1.4.15 Issues of concern – questions and answers

These are commonly asked questions relating to designers' duties under CDM.

1 **Q** To what lengths do designers need to go in considering hazard and risk and providing guidance text for pre-tender health and safety plans?

A Regulation 13(3) states that these issues need to be addressed –

"...to the extent that it is reasonable to expect the designer to address them at the time the design is prepared and to the extent that it is otherwise reasonably practicable to do so."

Any competent designer, therefore, has to be prepared to take a view on what is reasonable in the circumstances, all things being considered, and at that time. Properly informed professional judgements need to be made. What will not be acceptable is to fail to consider hazards.

2 **Q** When preparing input to the pre-tender health and safety plan, how does a designer decide whether residual risk has to be pointed out to a contractor?

A When the residual risk is of a site-specific or project-specific nature that a competent contractor might not recognise, it should be drawn to the attention of contractors.

3 **Q** How does a designer decide whether to mention the need for specific controls?

A This will be when:
(a) the client wishes a specific approach to be adopted
(b) the design requires specific precautions, or
(c) a significant risk might not be recognised by a competent contractor.

4 **Q** When considering how far to go in documenting project risks, does the likely type of contracting organisation need to be taken into account?

A This is a matter of judgement for designers based on what is known at the time, bearing in mind that they should be able to rely upon appointment of a competent principal contractor and other contractor(s). If the contractor has already been appointed, the best way forward is to agree with them the information that they need.

5 **Q** Can information be shown on a number of documents as well as in the pre-tender health and safety plan, and does it need to be duplicated?

A Specific concerns must be given in the pre-tender health and safety plan. Further information may be provided elsewhere (eg drawings, specifications and other documentation) and cross-referenced in the plan. If they are important, experience shows that they will be more likely to be acted on if shown on the drawings or referenced from the drawings. Information on drawings is most likely to be seen by the individuals directly engaged in and managing the work on site.

6 **Q** Are control measures recommended by designers binding upon the contractors?

A Not unless it is written into the contract.

7 **Q** How can a designer decide on levels of risk and balance options?

A Decisions must be based on knowledge and experience. Advice from colleagues or experts should be sought if there is uncertainty on a particular point. Contractors may also be able to advise the designer on the relative health and safety merits of different design options. Note that designers should not agonise between options that will present similar levels of risk, but concentrate on those that could substantially reduce it.

8 **Q** The ACoP (para 127) lists hazards that designers should *where possible* take particular actions to design out and/or reduce levels of risk; what does *where possible* mean?

A The HSE advises that designers should follow the advice unless there are strong reasons to do otherwise. Designers may be asked to explain their reasons.

9 **Q** Do the health and safety aspriations of a client affect decisions?

A If a client wishes to aim for a higher standard of health and safety than would normally be required for basic legal compliance, this may affect decision-making and the site rules in the pre-tender health and safety plan. The client's requirements in this respect should be made plain at an early stage. The basic requirements of CDM will not be affected.

10 Q Does quality assurance (QA) play a role in design risk assessment (DRA)?

A This is a business decision. However, DRA is fundamental to CDM and QA may be an appropriate management tool to monitor whether it is being carried out systematically and to demonstrate that the business is complying with statutory duties.

11 Q How do the contractors' risk assessments relate to DRA?

A DRA is used by designers to identify and assess hazards and to help them make decisions about design options. Contractors use risk assessment to develop a safe system of work that takes account of information in the health and safety plan and all project circumstances.

12 Q If the project is at an early stage (eg bidding for grants, obtaining estimates) or if no planning supervisor is appointed by the time a designer is making decisions, is a designer relieved of his duties under Regulation 13?

A No. Regulation 13 duties are not dependent on the appointment of a planning supervisor.

1.4.16 Getting help

The HSE runs a helpline, tel 08701 545 500 fax 02920 559 260 or email <hseinfomationservice@natbrit.com> you can also visit their website <www.hse.gov.uk>.

1.5 HOW TO USE THIS GUIDANCE

The work sectors have been organized into five groups (A to E), each of which addresses a major area of construction work. Within each group the work sectors are referenced numerically. The arrangements of the groups has been organized to reflect the order of construction where possible.

Each of the 39 work sectors is referenced as **<Group letter><Work sector number>** For example:

Deep basements and shafts is referenced as B2 (Group B, Work sector 2 in this group).

This referencing system is used for the internal cross-references between work sectors, principally in the Related issues section.

In some instances work sectors are intended to be read in conjunction with others, eg C1 (General concrete) should be read alongside any of the more specific concrete topics (C2,C3 or C4).

A detailed description of the way in which the work sectors are arranged is given in Chapter 2.

1.6 FLOWCHART OF HAZARD ANALYSIS

Designers can follow a logical process to deal with the hazards that have been identified, either as typical generic hazards or as project specific. The outcome of this process will depend on the nature of the hazards and the limits of the project that is being considered. Figure 1.1 illustrates the process as a flowchart, starting with the identification of the hazards and concluding with the significant residual risks.

The three elements of Figure 1.1 are described below.

1 The design process is divided into three stages (concept, scheme and detail). Each design stage will address different issues, but typically the nature of issues will move from the general to the particular as the design is refined.

2 Once hazards have been identified, the CDM process is split into the three approaches (avoid, reduce, control) and each of these methods of solution is considered in turn. For example, reduction measures may be considered only if a hazard cannot reasonably be avoided.

3 The end of the hazard consideration in design process is the consideration of the residual risks that remain in the design. Those that are judged to be significant or unusual in the context of the project need to be identified. Information on these risks may need to be provided to those who need it in order that appropriate action can be taken by the contractor, for construction, or the client (and others) for future work. This is considered further in Chapter 3.

This process of analyzing and dealing with hazards is considered in the work sectors under the headings of *Hazard consideration in design* and *Examples*. The Hazard consideration in design section within each work sector assumes a parallel series of operations to analyse the various risks arising from the intended construction. These then determine the degree of effort designers should apply to eliminating or reducing the hazards and then informing others so that they can control them. This total process is referred to as ERIC (Eliminate, Reduce, Inform, Control).

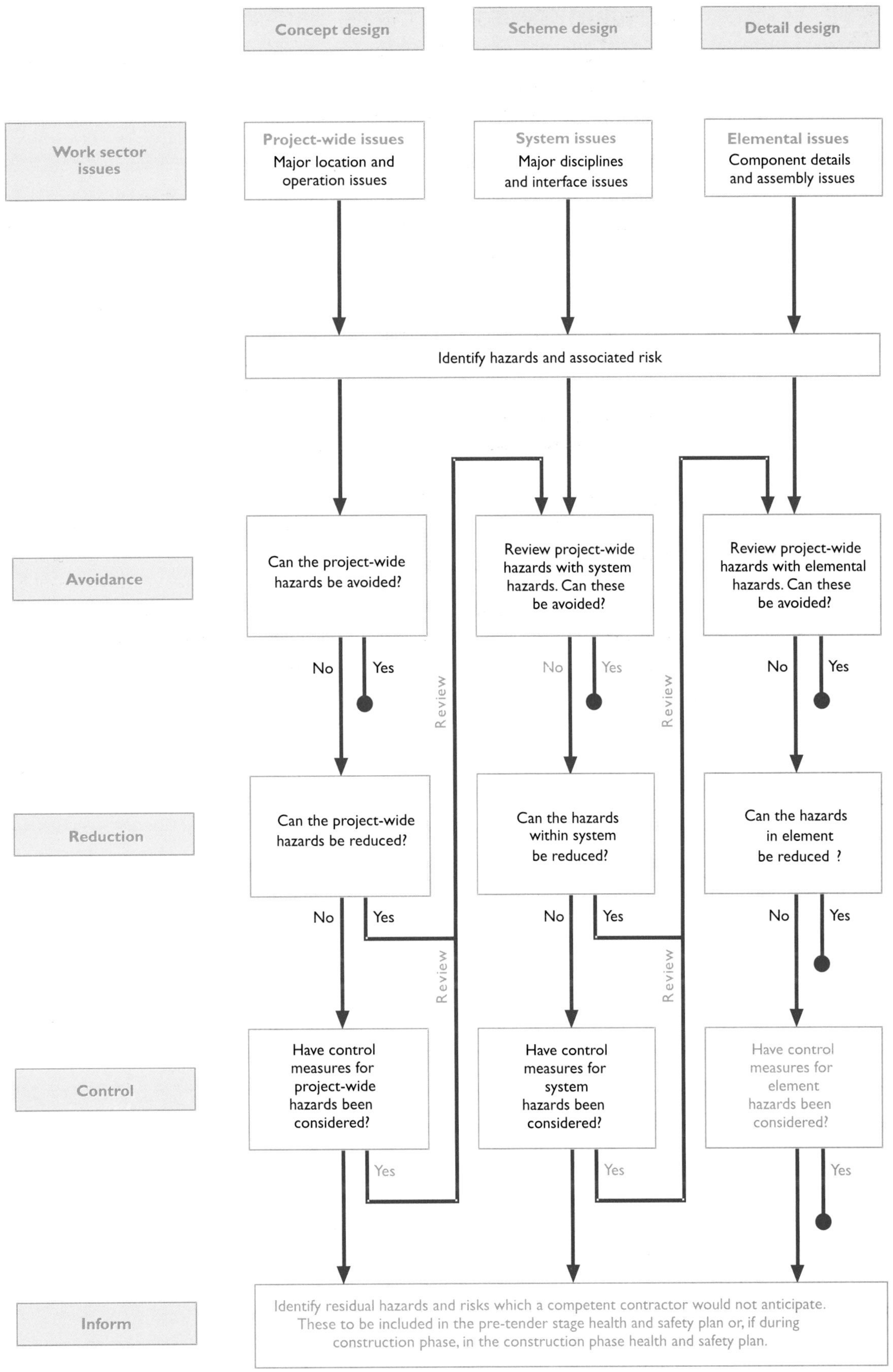

Figure 1.1 *Flowchart of hazard analysis*

1.7 GENERIC ISSUES

1.7.1 Introduction

The identification and consideration of health and safety hazards by work sector must be set within the wider context. This section looks at three such issues:

- common types of accident
- major health hazards
- planning and sequencing of construction.

There may be other issues that need to be considered on each particular project and designers should consider them in conjunction with the hazards related to work sectors.

1.7.2 Common types of accident

The main causes of injury in construction work are set out in Table 1.4. They happen in many different work sectors. It is essential for designers to have an appreciation of the part that they can play by eliminating the hazards and reducing the risks involved. Examples of preventative action are also shown in Table 1.4. It is important to understand that there are many designers involved in projects, including people who specify the materials to be used. Each needs to play their part, even though they cannot each address all of the issues.

Table 1.4 *Issues relating to types of accident*

Risk	Examples of causes	Examples of preventive action
Falls from height	Fall from a flat roof Fall through fragile roof/skylight Fall from ladder	Design in parapet or barrier Do not specify fragile materials Design out the need for ladders during construction, cleaning and maintenance operations eg: • sequence work to allow permanent stairways to be used during construction • specify/sequence work to provide hard standing for mobile access equipment • design windows to be cleaned from the inside • specify materials that don't need routine painting or design in safe access for maintenance • consider prefabrication so that sub-assemblies can be erected at ground level and then safely lifted into place Design in permanent access equipment eg: • rails for cradles/gantries • anchorages for fall arrest equipment and rope access equipment
Trench collapse	Poor ground conditions	Ensure that there is adequate information about ground conditions and ground water Seek to avoid deep excavation in poor conditions
Struck by moving vehicle	Poor site layout	Design site to: • provide safe access and egress onto public roads • ensure there is adequate space for plant and equipment to operate safely
Electric shock or burn	Contact with overhead or underground cables	Provide accurate information on all cables Arrange for service diversions ahead of main works Position structures to minimise risks from: • buried services • overhead cables
Collapse of structure during erection	Incorrect sequence of assembly leading to temporary instability	Suggest sequence of erection Design in bracing to ensure stability during erection

1.7.3 Major health hazards

Many more construction workers suffer serious health problems due to their work than serious accidents. Many of these health hazards can be avoided at the design stage. Table 1.5 gives some examples.

Table 1.5 *Issues relating to types of health hazard*

Health problem	Typical causes	Preventative action
Skeletal/muscular disorders	Laying heavy concrete blocks	Specify light blocks (n/e 20 kg)
	Laying kerb stones	Specify light block kerbs or extruded kerbs/ heavy sections that can only be lifted mechanically
	Needing to lift in an awkward way eg needing to twist and turn, particularly repeatedly	Design for ease of access eg avoiding need for awkward postures or twisting in plant rooms
Hand/arm vibration syndrome (HAVS)	Scabbling	Specify surface finishes that do not require scabbling
	Hand tunnelling	Minimise hand tunnelling eg specify size that will accommodate machinery
	Breaking out pile heads	Design piles so that mechanical pile head removal is possible; specify method
Dermatitis	Allergic reaction due to contact with wet cement	Design to use bulk supply pumped into position to minimise skin contact Ensure that good welfare facilities are specified Specify low chrome cement if skin contact is likely
Noise-induced hearing loss	Noisy machinery/ processes	Avoid the need for noisy processes such as mechanical breaking
Asbestosis, mesothelioma	Exposure to asbestos dust	Ensure that site has been surveyed to identify/locate any asbestos
Solvent exposure	Specification of solvent-based adhesives for large surfaces or in confined spaces	Specify adhesives with non-volatile solvents eg water-based

1.7.4 Hazard Identification

A list of hazards is provided below to assist in appreciating risks. **A list of this type can never be complete.**

falls
plant movement
collapse
services
impact
health hazards
collapse of excavation
flooding/fire/explosion
services/electrocution
falling from height
noise/vibration/dust
inadequate working space
inadequate working platform
working environment
inadequate access
access to height/depth
manoeuvring vehicles
lifting, lowering loads
handling
untidy, unsafe site
radiation
polluted, contaminated access
struck by mobile plant
static machinery
electrical shock/fire/explosion
impact crushing
poor visibility
buried/crushed/trapped
plant and machinery
intrusive occurrences
ground effects/movements
inundation
plant instability
working environment
structural instability
fumes/heat
irritants
crushing
discarded syringes and sharps
lasers
asbestos

1.7.5 Planning and sequencing of construction

At concept stage, the designers of a project can do a great deal to avoid and reduce significant hazards. One technique used to achieve this is to alter the way the construction is planned including the sequence of construction that is assumed. This is a powerful tool, but it requires wide understanding of the construction process and of the options that are feasible.

Examples

- Designers can reduce the potential for health problems arising from manual handling by producing a scheme that encourages mechanical lifting during maintenance operations and places a limit on the weights of objects to be handled manually.

- Designers can reduce the need to work at height when erecting a steel frame by designing the steelwork in modular sections, which can be pre-fabricated at ground level and sequentially lifted into place. This does not eliminate working at height entirely but should reduce it significantly. The safety of those who do have to work at height can further be enhanced if the ground slab is installed before the steel frame so as to provide a stable, flat working surface from which mobile elevating work platforms can be used for bolting up. This is much safer than using ladder access and quicker than scaffolding.

Although designers are not expected to specify particular construction methods or sequences, they will be expected to have considered possible alternatives when the hazards are being identified. If the assumptions of construction method and sequence become inextricably woven into the design, such that there is only one reasonable choice, then this will have to be made known.

Alternative methods and sequences will need to be considered to ensure that the appropriate choice is made. This will include factors other than health and safety and the final choice will reflect that health and safety is not expected to be pursued at any cost, rather that decisions have to be sensible and reasonable and balanced with other design decisions.

Table 1.6 demonstrates how options may be compared in a simple, effective manner, assisting decision-making.

Table 1.6 *Example of the assessment of alternative methods and sequences of construction*

Consideration	Option 1 – Assembly in-situ	Option 2 – Pre-assembly and lift into place
Health and safety	Operatives working at height, risk of people falling, risk of falling objects, risk of lifting many individual items	Fewer operatives working at height, risks from manipulating large modules, quicker site operation means shorter exposure times
Quality	Variable quality	Better opportunities for quality control
Programme	Assembly *in-situ* slower than prefabricating	Quicker than assembling *in-situ* in terms of site work. Pre-assembly can be elsewhere on site or remote. Requires longer lead time
Price	Usually little difference in cost of materials and components. Pre-assembly possibly more expensive in total manpower/double handling, eg needs larger crane, but should be more predictable out-turn price.	
Performance	In this instance, it is judged that there is no material difference in eventual performance of the structure arising from the choice of one construction method or another.	

Note that comparisons will not always be necessary; the ACoP (para 125) states that "There is little to be gained by detailed comparison of construction techniques, for example whether to specify a steel frame or concrete portal building. The focus should be on issues that are known to have the potential to reduce risks significantly…"

2 Work sector guidance

2.1 WORK SECTOR GROUPS

The work sectors in this report are divided into five groups, relating to main stages of construction work, as follows:

Group A	General planning
Group B	Excavation and foundations
Group C	Primary structure
Group D	Building elements and building services
Group E	Civil engineering

Within these groups, each of the 39 work sectors is presented in a standard format over four pages. Once readers are familiar with the layout and arrangement of the guidance material, it is expected that individual work sectors will be referred to on a case-by-case basis and the structure of the guidance has been designed with this in mind. However, it is vital that users have read and understood the introduction to the report and also have access to the list of general reference material contained in Chapter 4 of the report.

2.2 THE CONTENT OF EACH WORK SECTOR GROUP

Each of the work sectors is arranged over four pages, with a common structure of headings and subdivision of the information presented.

Table 2.1 indicates what each part of the common structure of the work sector guidance contains. In order to get maximum benefit from the guidance, readers must know where particular information is contained within each work sector.

> Manual handling is a hazard potentially applicable to all construction processes.
> It has not been repeated in each section but **must** always be borne in mind.

Table 2.1 *Description of contents of each work sector*

Page	Heading used	Contents
First	Scope	This describes what the work sector covers, either as a formal definition or as examples in bullet point form.
	Exclusions	This describes what is not included in the scope of the work sector, usually items that have been considered too specialist.
	Major hazards	These are hazards that will be most likely to arise when this work sector is included in a construction project; a maximum of six are given to concentrate on the major issues.
Second	Specific hazard Identification	This section gives typical issues that may arise, depending on the nature of the project the designer is undertaking, to add to those identified as major hazards. There are examples of specific questions, headed *Possible key considerations*. These are included to stimulate lateral thought, as it is impossible for any guidance of this kind to cover all possible eventualities. This list works in conjunction with the prompts.
	Prompts	This gives a list of general issues that may well influence how the key considerations are addressed. The list is NOT exhaustive and each prompt should be considered against each consideration.
	Hazard consideration in design	This section considers the issues that might arise at different stages of the design process (concept, scheme and detail), ranging from strategic questions at concept stage to specific issues at later stages. This section also presents some possible design options, which could allow the designer to avoid or reduce health and safety hazards. The design options are a starting point and must not be considered as a complete list.
Third	Examples of risk miitigation (methods of solution)	This sub-section considers typical issues, showing how these generate different responses, according to whether the hazard can be avoided or only reduced, or whether control measures are required as a last resort.
	Examples of risk mitigation (issues addressed at different stages)	This sub-section presents two typical projects to illustrate how a particular issue an be addressed in different ways as the design progresses from concept to detail stage.
Fourth	Related issues	This section identifies other work sectors within the report that have an impact on, or are affected by, the work sector in question. Work sectors are referenced by their letter and number – see Section 1.5.
	References for further guidance	This gives titles of other published guidance, arranged as primary, secondary and background references. General references that apply to the report as a whole are given in Chapter 4. The classification of a work sector in the Common Arrangement of Work Sectors, Civil Engineering Work Sectors and CI/SfB systems is also given where appropriate.

Group A – General planning
1 – Surrounding environment

Read the Introduction before using the following guidelines

Scope

- Infrastructure – roads, railways, waterways, airport – load, length, width, height requirements.
- Adjacent environment – residential, commercial, industrial, conservation, leisure, civil.
- Transportation – plant, materials, personnel, to/from site, emergency facilities
- Occupation and activities of existing environment.
- Working procedures in an occupied building, etc.
- Topography – mining, lakes, rivers, quarries, caves, reservoirs, tidal areas.
- Adjacent structures and services – tall structures, basements, foundations, pipes, overhead cables, road, rail.
- Air rights (oversailing), railway rights, navigation rights.
- Site access, availability – parking, storage, road closures, low bridges, special access for surveys to railways, rivers, etc.
- Security exclusion of public from sites.

Exclusions

- Special types of neighbourhood (eg nuclear installations, defence establishments).

Major hazards

Refer to the Introduction for details of accident types and health risks

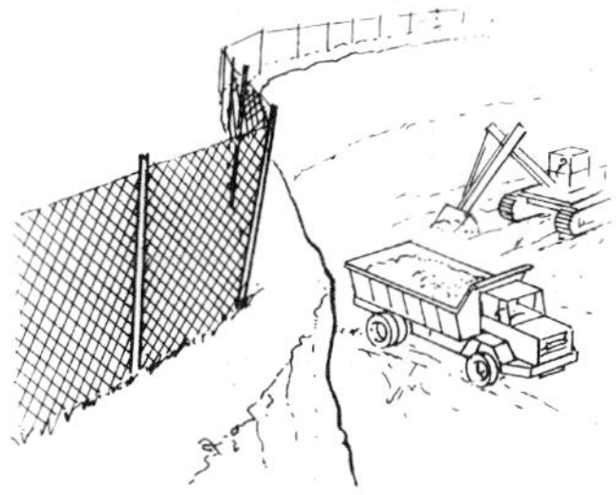

Falls
Work area perimeter security, preventing unauthorised access to active work areas.

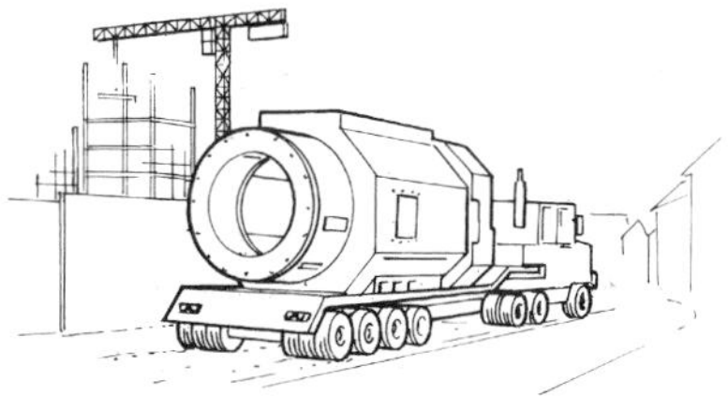

Plant movement
Onto and off site plant and transportation loads/access/routes/noise.

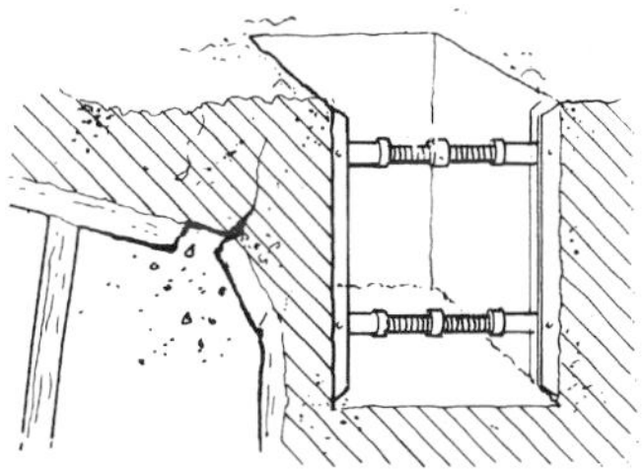

Collapse
Working adjacent to previous excavations or existing/adjacent structures, old mine workings.

Services
Existing services, overhead, underground, buried, unknown, on-site, in buildings.

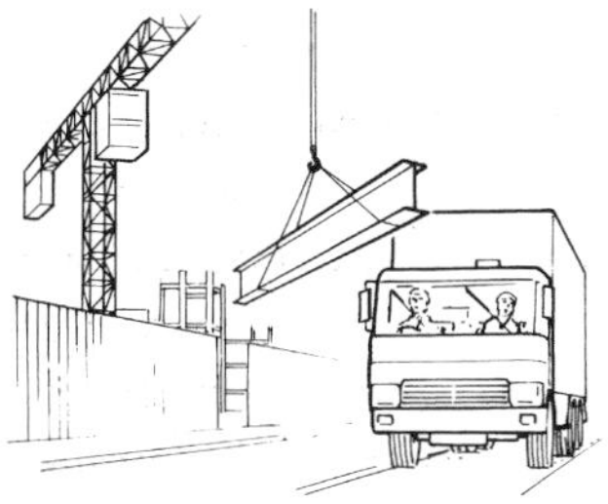

Impact
Plant lifting/over-sailing above local sensitive areas, shopping areas, schools, pavements, roads.

Health hazards
Dust, emissions from adjacent premises, dumps, sewers, ground seepage. Contaminated land.

A1 Group A – General planning
1 – Surrounding environment

Specific hazard identification

Possible key considerations:

- **Surroundings** – What are the health hazards in the proximity (airborne, buried, industrial, commercial liquid/dust/solid, etc)?
- **Infrastructure –** Where is the review of the infrastructure to ensure that it is able safely to support site traffic (load, frequency, duration, etc)?
- **Project** – What health hazards will the project construction place on the surrounding environment?
- **Methods** – Where have those situations that require method statements or controls been identified?
- **Information** – How will information be obtained from the client?
- **Services –** What hazards exist from surroundings (buried services, overhead cables, etc)?
- **Any others?**

Prompts

- Road infrastructure
- Rail/air/water transportation
- Services in site proximity
- Sequence of construction
- Ground conditions
- Sensitive areas/noise, vibration
- Offsite deposit
- Offsite storage
- Fire
- Emergency services

Hazards consideration in design

Stage	Considerations and issues	Possible design options to avoid or mitigate hazards identified from surrounding environment
Concept design	• Why build here? • Infrastructure scope • Neighbours assessed? • Shift working • Local climate • Access routes for site • Known contamination • Adjacent structures • Adjacent emissions • Oversailing adjacent buildings/areas • Pollutants • Local transport • Topography • Sensitive areas • External services • Existing structures • Security of site working • Vehicle movements • Sequence restrictions • Time restrictions	• Reduce contaminated ground hazards by containment. • Avoid open excavation to reduce hazards of adjacent contaminated ground. • Consider if the area poses increased risks of fire, flood or other hazards to site. • Ensure that local roads and infrastructure can carry the elements (as designed) without hazard to operatives/public.
Scheme design		• Design to suit local conditions, eg avoid basements in flooded ground; avoid work at height in exposed areas. • Use shafts and headings to avoid need for deep open trenches. • Foundations to suit site ground conditions (eg sheet piling in loose ground). • Avoid over-sailing adjacent property, fix windows from within building.
Detail design		• If shift or extended sequence of work required, design for off-site fabrication, minimise site activities. • For work in restricted areas, adapt the design (eg avoid brickwork in tidal excavations, use pre-cast elements – plastic or cast iron pipework). • Design to avoid existing services rather than re-route them.

Examples of risk mitigation (methods of solution)

ACTION \ ISSUE	Local traffic congestion	Damage to adjacent buildings and services
Avoidance Design to avoid identified hazards but beware of introducing others	Locate site access/egress away from local traffic.	Locate new foundations remote from adjacent buildings. Avoid locations with existing services.
Reduction Design to reduce identified hazards but beware of increasing others	Re-route local traffic away from site.	Allow for an independent compatible foundation. Divert existing services routes.
Control Design to provide acceptable safeguards for all remaining identified hazards	Control site traffic which is local to work activities.	

Examples of risk mitigation (issues addressed at different stages)

Old properties adjacent to the site are founded upon made-up ground. The basement of the new building is to cover part of the site and extends up to the adjacent party walls. The facade of the building on the site is to be retained, but is presently founded at a higher level than the proposed new building foundations.

Concept design	Scheme design	Detail design
To reduce the risk of collapse of adjacent structures resulting from ground movement, the designer has redesigned the new basement.	Allow the new structure to provide support to the facade during construction, avoiding temporary supports.	The designer will outline the procedures for supporting the retained facade during construction.

A new sewage treatment plant for a small town has been relocated, at feasibility study stage, due to surrounding ground conditions. However, the new location will require a longer route for the main sewer. The new sewer pipe has to pass through areas of existing services and has to withstand a high water table.

Concept design	Scheme design	Detail design
Designers are seeking ways of reducing costs of the increased length of main sewer, mindful of health and safety.	A design review identifies opportunities to reduce hazards of trench excavation through zones of existing services, by review of detailed alignment.	Access chambers will be designed in pre-cast concrete to include details that will reduce site activities in continuously pumped excavations.

Group A – General planning
1 – Surrounding environment

A 1 GENERAL PLANNING
B EXCAVATIONS AND FOUNDATIONS
C PRIMARY STRUCTURE
D BUILDING ELEMENTS AND BUILDING SERVICES
E CIVIL ENGINEERING

Related issues

References within this document

References	Related issues
A2	Site clearance activities will need to consider environment.
A3	Site investigation may confirm significant surrounding influences.
A4	Access onto the site is likely to be determined by surrounding environment.
A5	Site layout will be affected by the surroundings and access requirements.
B1–B3	Excavation planning needs to consider surroundings, underground, overhead services, etc.
B5–B7	Foundations may undermine or load surroundings.
C1–C4	Concrete materials brought to site in bulk.
C5–C6	Structural steel may require special access to site.
C7	Timber may be influenced by the surrounding environment.
C8	Masonry materials will require regular access to site for deliveries.
D5	Surface coatings/finishes may require special protection to prevent unsafe spreading during application.
D1	External cladding will need access and off-loading facilities.
D6	Cleaning buildings may oversail adjacent properties.
D7–D9	Services will be influenced by demands from site.
E2	Roads may affect access/egress limits to the site.
E3	Railways may assist or constrain the site access.

References for further guidance

Primary general references and background information given in Section 4

Primary

- CIRIA PR70 (1999) How much noise do you make? A guide to assessing and managing noise on construction sites
- CIRIA SP96 (1994) Environmental assessment – A guide to the identification, evaluation and mitigation of environmental issues in construction schemes
- HSE HSG66 Protection of workers and the general public during the development of contaminated land
- BRE Fire and explosion hazards associated with the redevelopment of contaminated land. Information Paper 2/87

Secondary

- DoT Safety at street works and road works: a code of practice 2001

Background

- CIRIA SP93 and SP94 (1993) Environmental issues in construction – A review of issues and initiatives relevant to the building, construction and related industries (volumes 1 and 2)
- CIRIA C512 (2000) Environmental handbook for building and civil engineering projects Part 1: design and specification
- CIRIA C528 (2000) Environmental handbook for building and civil engineering projects Part 2: construction phase
- CIRIA SP111 (1995) Remedial treatment of contaminated land Vol XI: Planning and management
- BS EN 10175 (2001) Code of practice for the identification of potentially contaminated land and its investigation
- Cairney, T and Hobson, C M (1998) Contaminated land: problems and solutions (Spon)

Classification

CAWS Group A, **CEWS** Class A, **CI/SfB** Code (--)(H)

Group A – General planning
2 – Site clearance and demolition

Read the Introduction before using the following guidelines

Scope

- Earth moving – loading, transporting, placing.
- Tree-felling, mast/pole/post.
- Waste removal – safe, toxic, suitable transport, licensed tip.
- Excavation – clays, granular, rock material, contaminated ground.
- Water removal – fresh, salt, foul, brackish, contamination of water courses.
- Services – above/below ground, tunnels, pipes, wires, diversions, relocations.
- Structures – above/below ground, party walls.
- Demolition – brickwork, timber, concrete, steel, asbestos, lead-based paints, etc.
- Partial removal of materials – site clearance/demolition.
- Recycling of materials – temporary storage, crushing, sorting

Exclusions

- Special conditions – unusual, unstable, (eg shallow mine workings).
- Radio-active areas.
- Below water – use of divers, dredging, etc.
- Buried explosives – mines, terrorist action.

Major hazards

Refer to the Introduction for details of accident types and health risks

Collapse of excavation
Buried obstructions, underground voids, unexpectedly poor ground conditions, water table.

Health hazards
Toxic waste, decomposing materials causing gaseous emissions, asbestos, microbiological agents.

Flooding/Fire/Explosion
From underground or above ground services.

Services/electrocution
Underground services, overhead cables, fixed equipment on site.

Falling from height
Loose/unstable material, access, inadequate support.

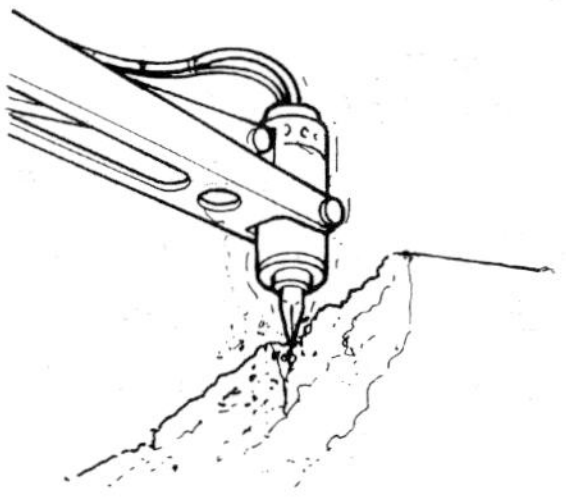

Noise/vibration/dust
Breaking-up of building fabric with mechanical breakers, asbestos, silica.

Group A – General planning
2 – Site clearance and demolition

Specific hazard identification

Possible key considerations:

- **Previous uses** – Where have previous uses of site been established?
- **Site survey** – When was a full site survey carried out?
- **Hazardous materials** – What are procedures to deal with hazardous materials?
- **Toxicity** – What is extent of hazardous materials' volume/toxicity, particularly asbestos?
- **Clearance** – Clearance causes dust hazards: prepare plans to define and contain it.
- **Unstable site** – Is the site unstable? Will clearance cause instability? Review any risks of flooding.
- **Demolition** – How are structures constructed and will it affect methods of demolition?
- **Handling** – What materials during clearance are hazardous to handle?
- **Services** – Identify existing above/below ground services and installations.
- **Any others?**

Prompts

- Existing health and safety file
- Client information
- Site records/drawings
- Investigation results
- Fill source
- Site security
- Site access
- Sequence of working
- Structural records
- Plant routes
- Personnel routes
- Health
- Fire

Hazards consideration in design

Stage	Considerations and issues	Possible design options to avoid or mitigate hazards identified for site clearance
Concept design	• Location – why here? • Bulk transport capability • Hazardous materials • Previous site use • Flood prevention • Demolition	• Ensure complete up-to-date description of existing site services. • Build remote from major existing services. • Avoid trees, rather than remove and replace.
Scheme design	• Phasing activities • Transport plan • Site contamination • Surveys/records known • Clearance methods • Demolition method • Elemental clearance	• Seal contaminated ground on site, if possible, rather than excavate and replace with fill. • Store and re-use material on site; avoid transportation off-site. • Avoid existing services rather than protect them. • Design service diversions to avoid/minimise site clearance. • Design new foundation to avoid existing obstructions.
Detail design	• Sequencing of work • Stability of ground • Stability of structures • Clearance activities • Demolition procedure • Landscape profile	• Optimise site filling with excavated material (cut and fill). • Use site demolition material as hard-core fill. • Use driven piled foundations through contaminated ground, to minimise open excavation.

Examples of risk mitigation (methods of solution)

ACTION / ISSUE	Contaminated ground	Demolition of large house on new development
Avoidance Design to avoid identified hazards but beware of introducing others	Proposed structures relocated, ground not disturbed.	Incorporate house into overall scheme with possible conversion to flats.
Reduction Design to reduce identified hazards but beware of increasing others	Contaminated ground identified and sealed.	Highlight site constraints.
Control Design to provide acceptable safeguards for all remaining identified hazards	Contaminated ground removed under controlled monitoring.	Recommend a sequence of demolition following survey.

Examples of risk mitigation (issues addressed at different stages)

A small enabling works contract is to be let for site clearance prior to the main contract. The site is adjacent to a main road which is particularly busy during rush hours. There are also colleges and schools in the area, so the pavements are usually crowded. The site clearance requires some ground works excavation and hardstanding. The site is to be fenced and made secure.

Concept design	Scheme design	Detail design
Careful consideration will be given to site layout and the surrounding environment, in order to reduce number of traffic movements.	Location of the site entrance and access roads will be planned to reduce hazards of site transport activities accessing and leaving the site.	The site entrance is designed to allow measures to be implemented to control site traffic across the pavement access.

Several properties dating from the late 19th century are to be demolished to allow for a new development. There are no construction drawings for the largest building on the site and the method of construction for this building will require to be established before demolition planning can take place. The building has been modified on several occasions since it was originally built.

Concept design	Scheme design	Detail design
The existing building will be surveyed in order to identify the structural systems and key demolition issues.	Part of the building is load-bearing brickwork with pre-stressed concrete slabs and beams. Information to be included in the pre-tender stage H&S Plan, to alert tendering contractors to demolition hazards.	To control site hazards of sudden failure, pre-stressed beams and slabs to be removed and broken up in a controlled condition at ground level. This is considered less hazardous than propping all the elements and breaking them up at height.

Group A – General planning
2 – Site clearance and demolition

Related issues

References within this document

References	Related issues
A4	Access onto/within the site to be kept clear and safe during site clearance.
A5	Site layout plays an important part in site clearance and following activities.
A1	Surrounding environment will need to be protected from the hazards of site clearance and demolition.
A3	Site investigation will be early and may need to interface with site clearance.
B1	General excavation could be part of or follow site clearance.
B2	Deep basement excavation could be required to interface with site clearance.
B3	Trenches for foundations/services for early enabling works.
B4	Retaining wall excavation may interface with site clearance.
B5	Ground stabilization may be needed before or after site clearance activities.
B6	Piling foundations will interface with site clearance and may form part of it.
B7	Foundation underpinning before or after site clearance.
E1	Small civil engineering works likely before or after site clearance .
E2	Work adjacent to roads needs to be considered together with all other public areas.
E3	Work adjacent to railways will have procedural implications on clearance and demolition.

References for further guidance

Primary general references and background information given in Section 4

Primary

- HSE HSG66 Protection of workers and the general public during the redevelopment of contaminated land
- HSE INDG258 Safe work in confined spaces
- HSE GS29 Parts 1 to 4 Health and safety in demolition work
- CIRIA SP78 (1991) Building on derelict land
- DoT, Interdepartmental Committee on the Redevelopment of Contaminated Land (ICRCL) Guidance Note 17/78 Notes on the development and after-use of landfill sites 1990

Secondary

- CIRIA SP38 (1985) The use of screens to reduce noise from sites
- CIRIA R97 (1992) Trenching practice
- HSE MDHS100 Surveying sampling and assessment of asbestos coating materials

Background

- BS 6031:1981 Code of practice for earthworks
- BS 6187:2000 Code of practice for demolition
- Institute of Demolition Engineers Orchard W.R. Elements of Risk and Safety Management in Demolition and Dismantling Projects
- Institute of Demolition Engineers Seminar Proceedings Demolition and the CDM Regulations, 2001

Classification

CAWS Group A&C, **CEWS** Class D,E,I,J,K,L,R,T,X&Y, **CI/SfB** Code (1-)(D2)

Group A – General planning
3 – Site investigation

A3

Read the Introduction before using the following guidelines

Scope

- Structural surveys – including those in confined spaces.
- Building services surveys.
- Materials survey (eg asbestos, lead-based paint).
- Ground/soil sampling.
- Topographical/ground movement surveys.
- Rock coring.
- Boreholes/deep drilling at small diameter.
- Trial pits and excavations.
- Field testing – in-situ tests.
- Existing services location/inspection – pipes, cables, sewers.

Exclusions

- Testing – eg underwater.
- Traffic surveys.
- Use of specialist equipment – towers, scaffolds, jetties etc.

Major hazards

Refer to the Introduction for details of accident types and health risks

Inadequate working space
Width, height for identified situations, entrapment, awkward heavy lifting, manoeuvring.

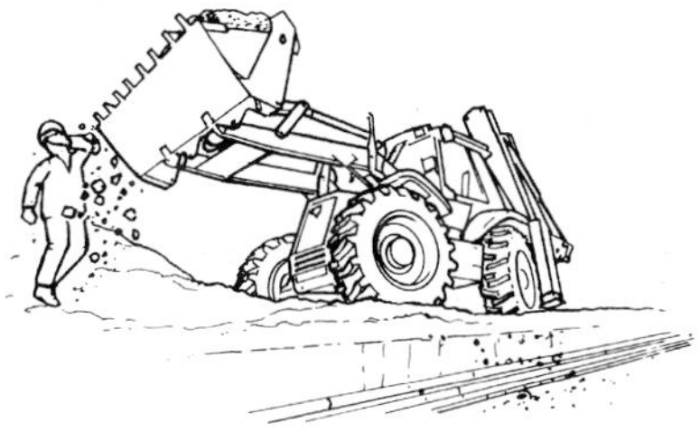

Inadequate working platform
Capacity to support load/weight, slope instability, underground voids.

Working environment
Ventilation, illumination, hearing, tripping, collision, falling through openings or from edges.

Excavation collapse
Depth/dimensions for soil conditions, water table shoring, supervision, descent, falling, collapse, slip.

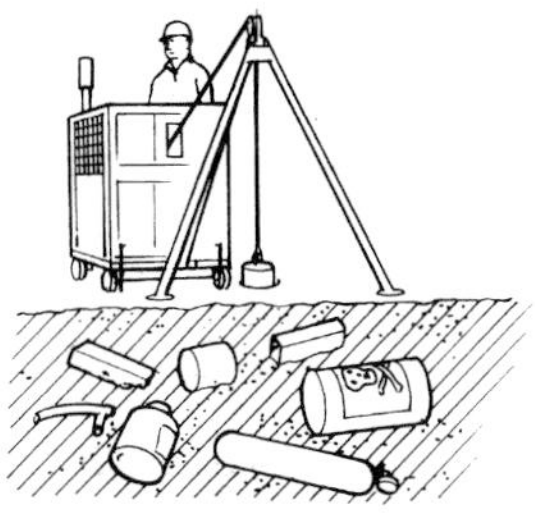

Health hazards
Noxious fumes, toxic, corrosive, explosive, microbiological, burial grounds.

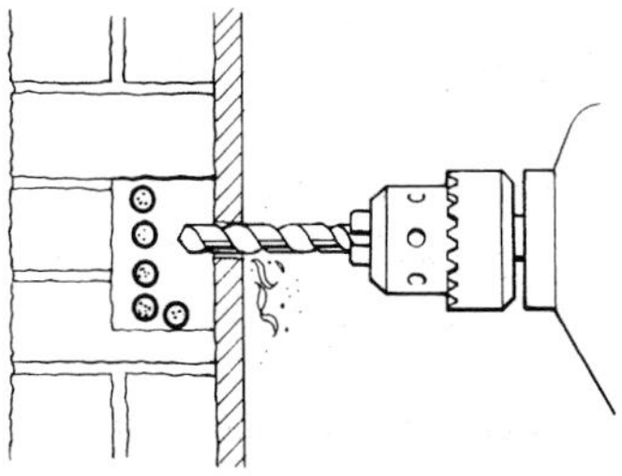

Services
Excavation, drilling.
Sampling/testing.
Location of existing services.

Group A – General planning
3 – Site investigation

Specific hazard identification

Possible key considerations:

- **Safe working** – Provision for working at height or in confined spaces or in traffic (etc)?
- **Route** – Where are adequate provisions for access to investigation point?
- **Access** – Plan adequate room for access around investigation zone?
- **Stability** – Have you ensured that the site or working platform is stable?
- **Asbestos and other contaminants** – Is full information available?
- **Contamination** – How is the contamination to be treated?
- **Services** – Where have services been located/exposed/disconnected?
- **Environment** – How will environmental control in confined spaces be done?
- **Equipment** – Where have you identified heavy equipment to be used?
- **Existing structures** – What effect will investigation have on existing structures and/or existing services?
- **What is the age and condition of buildings?**
- **Any others?**

Prompts

- Access routes
- Services
- Desk study
- Sequence of investigation
- Work access
- Site hoardings
- Health
- Fire
- Method statements

Hazards consideration in design

Stage	Considerations and issues	Possible design options to avoid or mitigate hazards identified for site investigation
Concept design	• Location – why here? • Previous use/surveys • Existing features • Access/exposure • Unknown features • Scope of survey • Stability of site • Working environment • Hazardous materials • Existing conditions • Survey equipment • Contamination • Setting out probes • Investigation access • Existing services • Temporary works • Specific conditions	• Check that sources of existing information will be able to provide relevant information. • Consider the likely completeness and accuracy of available information and the practicality of obtaining an independent survey or investigation.
Scheme design		• If weather has an influence on the intended investigation, check or define the preferred period of execution. • Review whether the investigation team could be better informed or protected by a pre-investigation survey or cleansing operation.
Detail design		• Relate the distribution of boreholes, etc to the location and size of potential voids/obstructions but with respect to accessibility and potential buried obstructions. • Highlight risks with the use of electricity in particular situations.

Examples of risk mitigation (methods of solution)

ACTION / ISSUE	Structural behaviour	Site exploration
Avoidance Design to avoid identified hazards but beware of introducing others	Previous structural drawings exist.	Boreholes samples and laboratory testing.
Reduction Design to reduce identified hazards but beware of increasing others	Pre-demolition structural survey.	Trial pits sampling.
Control Design to provide acceptable safeguards for all remaining identified hazards	Ensure technical demolition supervision.	In-situ field tests.

Examples of risk mitigation (issues addressed at different stages)

A soils site investigation is to be carried out. It is known that various live services cross the site. Measures will be required to be taken to avoid damaging live services and causing injury to operatives carrying out the site investigation.

Concept design	Scheme design	Detail design
Soils investigation, where possible, will be set out to avoid areas of known live services.	Exploration points are required in areas of existing services. To reduce the hazard, starter pits will be used for boreholes in close proximity to live services.	To reduce the hazard of live services, arrange for isolation of existing services not required, prior to starting site investigation work.

A group of buildings are to be extensively refurbished. Some drawings of the existing buildings are missing and existing services drawings are particularly incomplete. The services will be substantially replaced with a fully air-conditioned building.

Concept design	Scheme design	Detail design
A site investigation will be carried out to set-out existing services and insulation, to identify hazards from services and possible asbestos insulation.	During the survey to locate existing services, extensive asbestos insulation was found. The concept design was amended to avoid this hazard.	Designer sets down control procedures for isolating existing services and removing hazardous insulation prior to refurbishment activities.

Related issues

References within this document

References	Related issues
A2	Ground contamination will influence precautions required for spoil disposal
A2	Demolition strategy to be incorporated into brief for site investigation.
A4	Work in areas liable to flooding will generate special access requirements.
A4	Work on slopes may create stability problems and will generate special access requirements.
A5	Layout and structural capacity of existing buildings may influence feasibility and location of internal borings.
B1	Reinstatement of boreholes and pits to be considered in the design of any future excavation or foundation work.
B5	Shoring of trial pits will depend on stability of ground as assessed in the field.

References for further guidance

Primary general references and background information given in Section 4

Primary

- The Electricity Association, Avoidance of danger from underground electricity cables
- British Gas, Precautions to be taken when working in the vicinity of underground gas pipes
- HSE HSG47 Avoiding danger from underground services
- HSE INDG258 Safe work in confined spaces
- HSE HSG66 Protection of workers and the general public during the development of contaminated land
- CIRIA SP25 (1983) Site investigation manual (microfiche)
- CIRIA SP73 (1991) Roles and responsibility in site investigation
- CIRIA SP102–112 (1995/96) Remedial treatment for contaminated land – Vols 2–12

Secondary

- CIRIA R97 (1992) Trenching practice
- HSE GS6 Avoidance of danger from overhead electric power lines

Background

- BS 5930:1999 Code of Practice for Site Investigation
- Clayton C R I *et al* (1995) Site investigation (Blackwell)
- Institution of Civil Engineers (1993) Specification for ground investigation (Thomas Telford)

Classification

CAWS Group D, **CEWS** Class B, **CI/SfB** Code (1-)(A3s)

Group A – General planning
4 – Access (onto and within site)

A4

Read the Introduction before using the following guidelines

Scope

- Site perimeter – appropriate site access and security, boundary constraints.
- Site perimeter/building relationship – storage, handling and layout areas, welfare areas.
- Protection, illumination, visibility, ventilation, escape routes designed in.
- Building/structure, envelope, sequence of building operations.
- Emergency routes, compartmentation during construction, escape routes protection.
- Width, height, length, curvature, load – carrying capacity, overhead cables/services.
- Movement of personnel, vehicles, plant, materials.
- Site horizontal movement – roads, railways, waterways, footways, conveyors.
- Site vertical movement – stairs, ramps, shutes, hoists, cranes, pumps, cradles, lifts.

Exclusions

- Specialist access – aerial lifts, helicopters, abseiling, etc.
- Unstable structures – bomb-blasted, fire-damaged, defective, partial failure, etc.
- Tunnels and shafts.
- Radio-active contaminated areas.

Major hazards

Refer to the Introduction for details of accident types and health risks

Inadequate access
Width, height for identified situations. Capacity to withstand load/weight. Overhead cables.

Access to height/depth
Height, falling, struck. Depth; entrapment, asphyxiation, drowning.

Manoeuvring vehicles, plant and materials
Turning reversing, clearance, height, width. Two-way traffic, lack of storage layout.

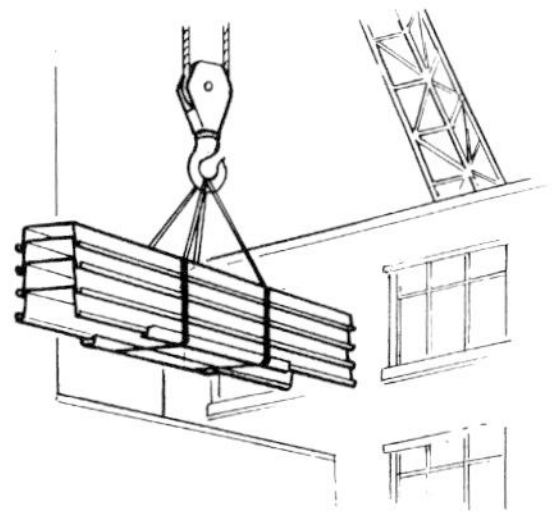

Lifting, lowering loads
Clearance, lifting, slewing, handling, visibility, site lines, capacity, oversailing public access areas.

Working environment
Noise, non-separation of plant and people, poor visibility.

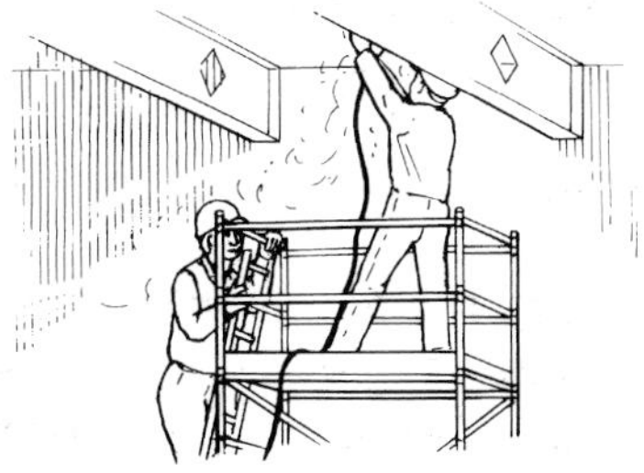

Polluted, contaminated access
Poor ventilation, poor illumination, dust, damp, dirt.

Group A – General planning
4 – Access (onto and within site)

Specific hazard identification

Possible key considerations:

- **Access** – Plan adequate provision for access at site perimeter?
- **Site traffic** – Where will site traffic cross a public road, pavement and/or existing services to gain access to the site?
- **Access zone** – Ensure there is adequate room for access around/over the construction zone within the site at all levels.
- **Sequence** – Where will access within the site have to change as construction progresses?
- **Constraints** – Does the client's on-going business affect the work area, eg supermarket, hotel, process plant?
- **Separation** – Set out separation of vehicles/plant and from operatives activities.
- **Interfaces** – Carefully identify vehicles/plant/operatives crossings/interfaces.
- **Lifting zone** – Where have you identified lifting zones (locate cranes/hoists)?
- **Site layout** – Where have you identified delivery, laydown and storage areas to reduce manual handling and allow for mechanical lifting appliances?
- **Any others?**

Prompts

- Public – pedestrian, vehicular
- Plant and machinery routes
- Pedestrian routes
- Manual handling
- Mechanical lifting
- Movement of large units
- Existing services – divert?
- Overhead cables – divert?
- Lifting zones
- Restrictions, weight, size
- Emergency services
- Access routes
- Materials routes
- Storage areas
- Work areas
- Welfare areas
- Health issues
- Sequence of construction
- Fire
- Security
- Signage

Hazards consideration in design

Stage	Considerations and issues	Possible design options to avoid or mitigate hazards identified for access (onto and within site)
Concept design	• Location – why here? • Main access, egress • Building activities • Emergency services • Site perimeter secure • Footprint of structure/building • Access within site and around building systems • Within perimeter • Movements in space between elements • Access in relation to site elements • Access to each building element	• Traffic study to improve layout of site entrance and access. • Ensure phased completion maintains construction access routes. • Identify adequate site access to allow the design to be built safely. • Design to maintain emergency access to site.
Scheme design		• Design of permanent works to allow adequate temporary site access. • Design to provide for safe access to repair and maintain the building. • If access is conditional upon sequence, define the necessary sequence for safety throughout. • Emergency – identify and display services isolation points.
Detail design		• Consider the likely working space required for safe construction of the various elements. • Consider the site work location (eg avoid site welding over/near public). • Consider light, ventilation and time available for the site activity. • In emergencies, workforce accessible to paramedics and stretchers.

Examples of risk mitigation (methods of solution)

ACTION \ ISSUE	Deliveries to city site building	Framed building construction
Avoidance Design to avoid identified hazards but beware of introducing others	Design to allow lorries to unload within building footprint.	Design for permanent stairways to be installed as part of frame.
Reduction Design to reduce identified hazards but beware of increasing others	Design components to be packaged to reduce unloading time in road.	Design fixings inside lift shafts to support temporary stairs.
Control Design to provide acceptable safeguards for all remaining identified hazards	Highlight need to plan deliveries at off peak periods.	Highlight hazards involved when using ladders.

Examples of risk mitigation (issues addressed at different stages)

A city development site has very restricted access. The design team have agreed that special consideration will have to be given throughout the design to the particular difficulties of accessing this project during construction and subsequent maintenance.

Concept design	Scheme design	Detail design
To avoid site access difficulties, elements to have overall envelope of not more than 10m length x 3m x 2m x 20 tonnes.	To reduce site access hazards of the large atrium roof spans, a review of the structural element option has identified steel roof trusses as the preferred option.	Detail design hazards will be avoided with a steel roof truss to be designed for final assembly at ground level on site and then lifted into position.

A hospital has been planned to have the plant room for major plant at basement level.

Concept design	Scheme design	Detail design
Provision will be made for 4 boilers in the plantroom at basement level. Only 3 boilers will be installed during construction, through temporary works access over the plantroom.	The design will allow for a dedicated access route for boiler maintenance, removal and installation of the 4th boiler.	Design load and space checks on dedicated plant access route to include boilers, trolleys, lifting beams and winch points.

A 4 GENERAL PLANNING
B EXCAVATIONS AND FOUNDATIONS
C PRIMARY STRUCTURE
D BUILDING ELEMENTS AND BUILDING SERVICES
E CIVIL ENGINEERING

Group A – General planning
4 – Access (onto and within site)

Related issues

References within this document

References	Related issues
A1	Surrounding environment will have an influence on the way that the site is accessed.
A2	Site clearance strategy to be incorporated into early phases of access planning of the site.
A5	Site layout of temporary and permanent works will have a considerable influence on the access plan, ensure sealing (fire protection) during construction.
B1-B4	Excavation may affect the phasing of access to the various work faces as well as the site perimeter.
B6	Piling planning should maintain adequate access to the site.
C5,C6	Structural steel erection will affect both horizontal and vertical movement, limiting access for other activities.
D1	Cladding delivery, storage, lifting, positioning and fixing require detailed consideration of access routes.
D4	Windows/glazing, because they are fragile require careful integration into the works sequence, perhaps with temporary dedicated access.
E2	Permanent road construction will need to be considered when sequencing the temporary/permanent site access plan.
E4,E5	Bridge construction/maintenance is likely to require specialised access to carry out the work.
E6	Work over water will generate access requirements that are equally, if not more, demanding than land-based work.

References for further guidance

Primary general references and background information given in Section 4

Primary

- CIRIA SP121 (1995) Temporary access to the workface: a handbook for young professionals
- HSE G47 Avoiding danger from underground services
- HSE INDG258 Safe work in confined spaces
- HSE GS6 Avoidance of danger from overhead electric power lines

Secondary

- CIRIA C703 (2003) Crane stability on site – second edition
- HSE CIS5 (revised) Temporarily suspended access cradles and platforms

Background

- CIRIA SP151 (2001) Site safety handbook

Classification

CAWS Group A, **CEWS** Class A, **CI/SfB** Code (--)(A5)

Group A – General planning
5 – Site layout

Read the Introduction before using the following guidelines

Scope

- Structure mass/site plan – restrictions imposed by adjacent buildings/land use.
- Site topography – alignment, stability, level, slope, terrain, groundwater.
- Site ergonomics – traffic routes, movements, personnel, plant, machinery, materials.
- Site entrances/exits – pedestrian, vehicular, site lines, site signage.
- Site security, prevention of flooding, interface with occupiers, other trades.
- Services, fuel storage – electricity, water oil, gas, overhead/underground.
- Plant layout/routes – tower cranes, mobiles/position, reach.
- Site accommodation – location of, and access to welfare, site storage layout.
- Emergency access/first aid – accommodation, facilities, staff.

Exclusions

- Specialist areas
  - Nuclear installations
  - Floating facilities
  - Petro-chemical works
  - Quarries
  - Airfields (Active).

Major hazards

Refer to the Introduction for details of accident types and health risks

Struck by mobile plant
Impact, crushing, slipping, falling, two way traffic, working cranes.

Static machinery
Inadequate protection and alerts to static machines operating.

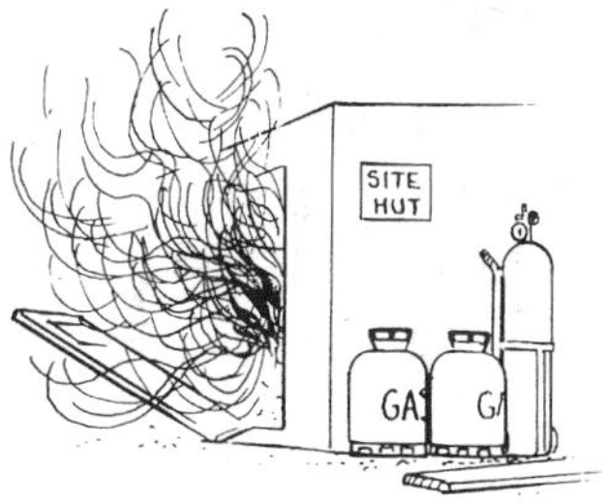

Electrical shock, fire, explosion
Location of site services, temporary, permanent storage of flammable gases, liquids.

Falling from height
Falling from structure or falling into excavation.

Impact, crushing
Objects falling from height from insecure location, work overhead, delivery and storage, unstable structures.

Poor visibility
Layout of access, roads, accommodation work area, people and material movers.

Group A – General planning
5 – Site layout

Specific hazard identification

Possible considerations:

- **Sequence** – What is the proposed sequence of construction?
- **Layout** – Ensure site layout proposals reflect the construction sequence. Has it been reviewed since last process change?
- **Phased access** – Is there adequate access onto site throughout all phases of the work?
- **Interfaces** – Where have access routes for plant and personnel been defined for each phase? Separation?
- **Services protection** – Define routes and protection for temporary and permanent site services.
- **Site storage** – Where is there adequate allowance for site storage?
- **Welfare** – Where is there space for welfare facilities and site offices?
- **Cranage** – What are the proposals for location of cranes?
- **Security** – What is site work area security during work activities for operatives?
- **Any others?**

Prompts

- Plant and machinery
- Vehicle/Pedestrians
- Sequence of construction
- Storage
- Welfare accommodation
- Access – temporary/permanent
  – vertical/horizontal
- Services (temp/permanent)
- Cranage (vertical movement)
- Conveyance (horizontal movement)
- Site access roads
- Temporary works
- Control measures
- Emergency services
- Fire prevention
- Fire-fighting
- Adjacent occupiers

Hazards consideration in design

Stage	Considerations and issues	Possible design options to avoid or mitigate hazards identified for site layout
Concept design	• Why this layout? • Vehicle/Pedestrian Separation • Site ergonomics • Block location • Project phasing • Emergency provision • Major movements • Facilities locations • System layout	• Is the site and/or its layout conducive for: – mechanical or manual construction? – off-site fabrication rather than insitu? – off site storage preferred to on-site? • Ensure the site layout can accommodate labour and plant necessary to complete the work on time. • Site access roads – one-way roads safer. • Prepare site layouts to reduce construction activities on site and therefore reduce risks.
Scheme design	• System phasing • Welfare facilities • Deliveries/access • Storage layout • Handling methods • Major services • Element layout	• Design of phased completion to allow continued safe construction and occupation of finished work, if required. • Consider width, height, load constraints for access to the work face. • Having decided on site layout for construction access, ensure maintenance access is also adequate.
Detail design	• Working space • Trade separation • Routing services • Positioning element • Identify services • Surveying	• Design to allow use of permanent roads as site roads. • Ensure minimum number of handling activities created by the design. • Avoid design sequence causing different trades to have to work together. • Design service routes to avoid active construction.

Examples of risk mitigation (methods of solution)

ACTION / ISSUE	Site layout logic	Traffic movement
Avoidance Design to avoid identified hazards but beware of introducing others	Plan site strategy/ergonomics. Phasing of site layouts with phased completion.	Road layout. Location on site.
Reduction Design to reduce identified hazards but beware of increasing others	Sequence each construction discipline.	Transportation study. Site road layout to be considered.
Control Design to provide acceptable safeguards for all remaining identified hazards	Consider site disciplines necessary. Identify site hazards created.	Site traffic procedures.

Examples of risk mitigation (issues addressed at different stages)

A terraced building in a city street is to be demolished and rebuilt. There are extensive parking restrictions in the area and the contractor will not be able to off-load materials or plant outside the site perimeter. There is no space available for unloading between the site perimeter and the proposed building.

Concept design	Scheme design	Detail design
Temporary vehicle access will be designed into the new building as part of temporary works to reduce hazards.	The design will improve access by omitting part of the first floor during construction, allowing off-street delivery and unloading.	Omit two bays of structure at first floor level, with ground floor structure able to carry site vehicles as temporary access area.

A 1960s office block is to be extensively refurbished. The existing heating systems are to be replaced with full air-conditioning and the concrete cladding units are to be replaced with insulated aluminium and glass cladding. The office must be kept operational as far as possible, consistent with the extensive refurbishment.

Concept design	Scheme design	Detail design
Options reviewed indicate hazard reduction with a full structural scaffold to carry out building refurbishment works, despite the restriction the scaffold imposes on site layout.	The facade scaffold layout will be designed to be amended to follow the refurbishment programme of activities and provide temporary support for the replacement of cladding units.	Design of structural scaffold to span access into the building loading bays for delivery vehicles to avoid unloading in the site area.

Group A – General planning
5 – Site layout

Related issues

References within this document

References	Related issues
A1	Surrounding environment could influence decisions on site layout.
A2	Site clearance will facilitate site layout planning.
A4	Access onto the site will be an integral part of the plan for the site layout.
B1-B3	Excavation will cause sequential implementation of site layout.
B4-7	Foundations as an early activity will determine site layout.
C1-C4	The supply of concrete and the installation of concrete structures will determine site layout.
C5-C6	Structural steel members, if significant in size and quantity will affect site layout.
D1	External cladding delivery and installation will interact with site layout.
E2	Roads integral/adjacent to the site will have a determining effect on site layout.
E3	Railways and their safety requirements are likely to affect site layout.

References for further guidance

Primary general references and background information given in Section 4

Primary

- HSE G47 Avoiding danger from underground services
- HSE INDG258 Safe work in confined spaces
- HSE HSG141 Electrical safety on construction sites
- HSE GS6 Avoidance of danger from overhead electric power lines

Secondary

- Building Employers Confederation (1995) Fire prevention on construction sites; Joint Code of Practice
- HSE G32 Safety in falsework for in-situ beams and slabs
- HSE HSG66 Protection of workers and the general public during development on contaminated land
- HSE INDG148 Reversing Vehicles
- CIRIA SP78 (1991) Building on derelict land
- CIRIA SP57 (1988) Handling of materials on site
- CIRIA PR70 (1999) Guide to assessing and managing noise on construction sites
- DoT, Interdepartmental committee on the redevelopment of contaminated Land (ICRCL) Guidance note 17/78 Notes on the development and after-use of land-fill sites 1990

Background

- CIRIA SP38 (1985) Simple noise screens for site use

Classification

CAWS Group A, **CEWS** Class A, **CI/SfB** Code (--)(A5)

Group B – Excavations and foundations
1 – General excavation

Read the Introduction before using the following guidelines

Scope

Excavations for:

- Foundations – buildings structures (bridges, retaining walls).
- Trenches – services, drainage, walls.
- Earthwork cuttings – road works, railways, services, water courses, minerals.
- Basements/shafts – buildings, drainage, services.
- Pits – ponds, waste disposal, archaeology.
- Heritage sites.

Exclusions

- Cofferdams and caissons – work adjacent to deep water, working in compressed air (see E7).
- Tunnels.
- Unexploded bombs.

Major hazards

Refer to the Introduction for details of accident types and health risks

Buried/crushed/trapped
Collapse of insufficiently supported sides of excavation. Undermining adjacent structures.

Falls
People, objects, supports, plant/machinery, materials. Unauthorised access.

Plant and machinery
Crushed/trapped/hit by plant falling/running into excavation. Fumes (eg carbon monoxide).

Intrusive occurrences
Groundwater, sewage, flooding.
Fire, explosion, smoke, gas leakage, methane

Health hazards
Contaminated ground/water.
Toxic chemicals.
Asbestos.
Microbiological diseases.
Dust and allergens.
Skin irritants.

Services
Electrocution.
Gas pipelines.
Sewage and process effluents.

Group B – Excavations and foundations
1 – General excavation

Specific hazard identification

Possible key considerations:

- **Site survey** - What are site conditions?
- **Ground Investigation** - Water table and slope stability?
- **Surface water** - What is local topography? Where are rivers, lakes, canals, ponds, etc.?
- **Working space** - What plant is required for backfilling?
- **Stockpile/disposal** - What are length and location of temporary roads and traffic routes to tip?
- **Adjacent property** - What are the position, type and use?
- **Services** - Where are they located?
- **Any others?**

Prompts

- Historical records
- Archaeology
- Contaminated land
- Waterborne diseases
- Access/egress
- Existing outfalls and vents
- Environment: general public, weather, buildings
- Working procedures

Hazards consideration in design

Stage	Considerations and issues	Possible design options to avoid or mitigate hazards identified for general excavation
Concept design	• Why excavate? • Structure location and layout • Site history • Materials • Ground investigation • Adjacent structure details • Watercourses • Disposal of excavated materials • Services • Slope stability • Working space • Waste disposal/haul routes • Stabilise excavation • Drainage	• An investigation of local earthworks could identify suitable minimum excavation site slopes. • Design to avoid possibility of excavation undermining adjacent buildings. • Consider possible locations of hard routes and tip sites to have least effect on construction and local community. • Ensure that ground investigation will detect likely site-specific contamination. • The ground investigation should relate to a range of option designs for the proposed structure.
Scheme design		• In low-lying waterlogged ground, consider dewatering methods to stabilise excavations. • For sites with limited working space, consider excavation within piled supports. • Assess risk of diverting services versus alternative of working around them.
Detail design		• For bored pile side supports in contaminated ground, consider handling and disposal of spoil. • Consider shoring options for adjacent structure, to maximise working space within site excavation. • Consider surface water drainage system to mitigate flooding of excavation.

Examples of risk mitigation (methods of solution)

ACTION \ ISSUE	Flooding	Unstable ground
Avoidance Design to avoid identified hazards but beware of introducing others	Raise level of building.	Stabilise ground.
Reduction Design to reduce identified hazards but beware of increasing others	Change foundation design from pads to piles.	Minimise depth of excavation. Batter side slopes.
Control Design to provide acceptable safeguards for all remaining identified hazards	Consider grout curtain, well-pointing or other drainage.	Outline appropriate propping requirements.

Examples of risk mitigation (issues addressed at different stages)

Buildings are to be demolished within an urban environment to provide a new shopping arcade with basement car parking. Neighbouring buildings abut the development site. There must be agreed and acceptable methods of work to prevent the collapse of excavations and buildings.

Concept design	Scheme design	Detail design
Identify history of site. Provide fully engineered retaining wall(s).	Consider new/existing structure interface and possible demolition/excavation techniques. Stability of excavation paramount. Restrict horizontal wall movements.	Retaining walls left in place for the underground car park. Consider additional methods of wall stabilisation, permanent or temporary. Consider access needs for construction and of local populace during the works.

A proposed leisure centre is to be constructed on stable made-up ground. The proposed excavation for foundations and services could be in contaminated ground.

Concept design	Scheme design	Detail design
Obtain historical records/surveys of site, detailing extent of contamination, from client. Avoid excavation in areas of known toxicity. Locate services in areas where there is no contamination, if possible.	Avoid contact with contaminants. Reduce contact by design of capping layer or use of driven piles to minimise ground disturbance. Test for types of waste product.	Consider implications for unavoidable works in contaminated area and provide relevant information to the Planning Supervisor. Put services in ducts to minimise future ground disturbance.

Group B – Excavations and foundations
1 – General excavation

A GENERAL PLANNING
B 1 EXCAVATIONS AND FOUNDATIONS
C PRIMARY STRUCTURE
D BUILDING ELEMENTS AND BUILDING SERVICES
E CIVIL ENGINEERING

Related issues

References within this document

References	Related issues
A1	Local environment should be protected from the hazards associated with excavations. OH + UG services may need diversion, temporary support or protection. Adjacent structures (including walls) may need temporary support, additional support (shoring), underpinning, etc.
A2	Demolition could precede excavation. Locate the health and safety file.
A3	Ground investigations should include testing for toxic materials, soil density, gasses, stability and groundwater details.
A4	Access/egress routes should be suitable for emergency services and for assumed construction plant.
A5	Site layout should be related to the working space available, the excavation plant to be used and the size of the site.
B2	- deep basements and shafts
B3	- trenches for foundations and services
B4	- retaining walls
E1	Working space availability on a small site accentuates the hazards of excavation. Careful planning is required.
E2,E3	Excavation adjacent to roads and railways should include for safe clearance distances.
E6,E7	Working near water introduces a greater risk that the excavation may flood.
	For issues associated with specific excavation work, refer to other sections in this group, eg:

References for further guidance

Primary general references and background information given in Section 4

Primary

- CIRIA R97 (1992) Trenching practice
- CIRIA R130 (1993) Methane: its occurrence and hazards in construction
- HSE HSG47 Avoiding danger from underground services
- HSE HSG66 Protection of workers and the general public during the development of contaminated land
- HSE CIS8 (revised) Safety in excavations
- HSE INDG258 Safe work in confined spaces

Secondary

- HSE GS6 Avoidance of danger from overhead electric power lines
- HSE GS29 Parts 1 to 4 Health and safety in demolition work
- CIRIA C515 (2000) Groundwater control – design and practice
- CIRIA SP32 (1984) Construction over abandoned mineworkings
- CIRIA PR77 (2000) Prop loads in large braced excavations

Background

- BS 6031:1981 Code of practice for earthworks
- BS 8002:1994 Earth retaining structures
- Trenter, N A (2001) Earthworks: a guide (Thomas Telford)

Classification

CAWS Group D, **CEWS** Class E, **CI/SfB** Code (1-)C

Group B – Excavations and foundations
2 – Deep basements and shafts

B2

Read the Introduction before using the following guidelines

Scope

- Basements for buildings, multi-storey offices, residential block, warehouses, car parks, etc.
- Large diameter access shafts (diameters greater than 3 metres).

Exclusions

- Piles up to 3 metres diameter (see B6).
- Shafts for mining industry.
- Cofferdams (see E7)

Major hazards

Refer to the Introduction for details of accident types and health risks

Falls
Falling into excavation or shaft. Debris or materials dropped into shaft or excavation.

Buried/crushed/trapped
Excavation collapse due to lack of or insufficient support.

Health hazards
Contaminated ground and water.

Services
Gas pipelines.
Electrocution.
Flooding by water or effluent.

Ground effects
Sudden inrush or ground water.
Mudslides.
Artesian boiling.
Methane.

Plant and machinery
Crushed/trapped/hit by plant in excavation.
Liquid fuels and exhaust fumes.

B2 Group B – Excavations and foundations
2 – Deep basements and shafts

Specific hazard identification

Possible key considerations:

- **Site survey** - Where are adjacent structures and services located?
- **Ground investigation** - What condition is soil, what is water table level?
- **Hydrology** - Where are water courses and aquifers?
- **Gases** - What level of ventilation is required?
- **Working space** - What is required for working ramps, stockpiling, deliveries of machinery?
- **Supports** - What systems will be appropriate?
- **Technological developments** - How can these be incorporated?
- **Any others?**

Prompts

- Historical records
- Archaeology
- Contaminated land
- Waterborne diseases
- Access/egress
- Existing outfalls and vents
- Environment: general public, weather, buildings
- Working procedures

Hazards consideration in design

Stage	Considerations and issues	Possible design options to avoid or mitigate hazards identified for deep basements and shafts
Concept design	• Why excavate? • Structure location and layout • Site history • Materials • Ground investigation • Adjacent structure details • Watercourses • Disposal of excavated materials • Services • Slope stability • Working space • Waste disposal/haul routes • Stabilise excavation • Drainage	• On the basis of site investigation, determine excavation methods to be excluded and identify hazardous materials and ground water problems to be addressed. • Consider excavation proposals to minimise the effects on overhead and underground services or divert services either temporarily or permanently. • Consider excavation within contiguous bored piles which will form permanent walls (eg for a basement). • Ensure ground investigation confirms the extent and type of any contamination.
Scheme design		• Ensure ground investigation confirms the extent and type of any contamination. • Excavation must not compromise the structural integrity of adjacent buildings. (Consider effects on ground stability.) • Design to take into account working space available and effects on local environment.
Detail design		• Support details should be sufficient to contain or exclude contaminants. • For excluded material, consider safe methods of disposal. • Appropriate sharing techniques of adjacent buildings need to be identified (eg ground anchors). • Design to identify appropriate practical and safe temporary access. • Designer's assumptions on construction, sequence and methods to be identified in health and safety plan. • Organise advance diversion of services by the service provider.

Examples of risk mitigation (methods of solution)

ACTION \ ISSUE	Collapse	Methane from the ground
Avoidance Design to avoid identified hazards but beware of introducing others	Design for shaft construction and inspection using remote control methods (eg robotics).	Incorporation of a gas – impermeable membrane is the first construction activity.
Reduction Design to reduce identified hazards but beware of increasing others	Assume mechanical excavation only, with operator in safety cab at all times.	Install continuous mechanical ventilation.
Control Design to provide acceptable safeguards for all remaining identified hazards	Manual operations to be easy and of short duration, within the protected area.	Ventilation system activated by gas sensor

Examples of risk mitigation (issues addressed at different stages)

Part of a railway terminus is to be sold for property development.
The site contains an abandoned shaft and connecting tunnels which must be isolated from tunnels still in use.

Concept design	Scheme design	Detail design
Consider condition of shaft and possible re-use, as alternatives to backfilling.	Does isolation require local breakage and removal of the tunnel linings, or will substantial "plugs" be sufficient?	Check that there is a sequence of filling which keeps escape routes open and minimises depth of uncured/fluid backfill material.

A bridge construction requires the sinking of vertical shafts to complete the pier foundations. Ground investigation has revealed the presence of hard granite boulders which precludes boring techniques.

Concept design	Scheme design	Detail design
Consider possible and appropriate mechanical and/or manual excavation techniques.	Consider appropriate excavation techniques, related to location, access, topography, ground stability, plant.	Materials disposal, temporary access, stockpile. Protection of personnel in hole - support design.

Group B – Excavations and foundations
2 – Deep basements and shafts

Related issues

References within this document

References	Related issues
A1	Local environment should be protected from the hazards associated with excavations. In particular, adjacent structures (including walls) may need temporary support, additional support (shoring), underpinning, etc. Consider also movements of local populace and location of services/utilities.
A3	Ground investigations should include testing for toxic materials from former use, soil density, gases, stability and ground water details.
A4	Access/egress routes should be kept clean, clear of obstructions and be suitable for emergency services.
A5	Site layout should be related to the working space available, the excavation plant to be used and the size of the site.
E1	Working space on small sites accentuate the hazards of excavation.
E2,E3	Excavation adjacent to roads and railways should include safe clearance distances and/or controls for site personnel together with control of excavation traffic.
E6,E7	Working near water introduces a greater risk that the excavation may flood.

References for further guidance

Primary general references and background information given in Section 4

Primary

- HSE HSG47 Avoiding danger from underground services
- HSE HSG66 Protection of workers and the general public during the development of contaminated land
- HSE INDG258 Safe work in confined spaces
- HSE GS6 Avoidance of danger from overhead electric power lines
- HSE CIS8 (revised) Safety in excavations
- BS 8008:1996 Guide to safety precautions and procedures for the construction and descent of machine-bored shafts for piling and other purposes

Secondary

- CIRIA R139 (1995)Water resisting basements
- CIRIA C517 (1999) Temporary propping of deep excavations – guidance on design

Background

- The Institution of Structural Engineers Report on Design and construction of deep basements 2nd edition 1981
- Puller M, Deep Excavations: a practical manual (Thomas Telford)
- Tomlinson M J, Foundation design and construction (Prentice-Hall)
- BS 8004:1986 Code of practice for foundations
- CIRIA C580 (2003) Embedded retaining walls – guidance for economic design

Classification

CAWS Group G, **CI/SfB** Code (1-)C

Group B – Excavations and foundations
3 – Trenches for foundations and services

B3

Read the Introduction before using the following guidelines

Scope

- Trenches for strip foundations and pad foundations.
- Trenches for public utilities, services, pipe lines, drains, etc.
- Pits and pipe jacking.

Exclusions

- Deep pier foundations, underpinning.
- Trench depths > 6m and below water.
- Trenches adjacent to deep water.

Major hazards

Refer to the Introduction for details of accident types and health risks

Services
Electrocution.
Gas pipelines.
Sewage and process effluents.

Health hazards
Water-borne toxins and diseases.
Methane.
Asbestos.

Inundation
Flooding, drowning.
Chemical waste, effluents.

Plant and machinery
Vehicles too close cause trench side to collapse.
Fuel vapours and exhaust fumes.
Pneumatic tools (vibration white finger).

Falls
No guard rails, congested sites, lack of working space.

Buried/crushed/trapped
Materials/spoil, falls into trench. Trench collapse due to insufficient support.

Group B – Excavations and foundations
3 – Trenches for foundations and services

Specific hazard identification

Possible key considerations:

- **Site survey** - Where are the services?
- **Services** - Do services need support across trench?
- **General investigation** - What are soil types? What is the water table level? Could the trench flood?
- **Hydrology** - Where are the water courses? Will surface water run-off be a problem?
- **Side supports** - What systems are appropriate for the known ground conditions?
- **Gases** - What assessment is required to decide ventilation?
- **Working space** - What space is required for deliveries, stockpiling, ramps, haul road? Ensure safe distance from trench.
- **Backfilling** - What compaction can be carried out without using hand-held machines?
- **Any others?**

Prompts

- Historical records
- Archaeology (where appropriate)
- Contaminated land
- Protection of public
- Access/egress
- Illumination
- Waste disposal
- Noise/vibration
- Emergency procedures
- Propriety systems
- Confined space working
- Pipe jacking
- Trenchless methods

Hazards consideration in design

Stage	Considerations and issues	Possible design options to avoid or mitigate hazards identified for trenches for foundations and services
Concept design	• Why trenches? • Site inspection • Site history • Access/egress • Working space • Environment • Topography • Services • Ground conditions • Contamination • Rights of way • Plant/machinery • Trench supports • Drainage proposals • Adjacent structures • Working procedures	• Establish details of foundations of existing structures and previous work at the site. • Identify necessary working space in relation to existing buildings and services. • Identify (and reduce) effects on local environment and services. • Look for alternative routes to avoid deep trenches and steeply sloping ground. • Establish ground conditions
Scheme design		• Identify safe working space for excavation depths required. • The presence of water can create instability. Minimise need and depth of excavation. • Consider construction techniques to avoid need for trench excavation (eg local excavation and pipe jacking or slit trench and automatic cable laying).
Detail design		• To minimise effects of water, include for drainage system to restrict/divert ground waste and run-off. • Consider problems of existing services which cross proposed excavation. Identify necessary support systems. • Minimise the time and complexity of manual operations (eg jointing) to be carried out in trench.

Examples of risk mitigation (methods of solution)

ACTION \ ISSUE	Collapse of adjacent foundations	Flooding from adjacent watercourse
Avoidance Design to avoid identified hazards but beware of introducing others	Relocate or use shallower trench or use trenchless construction.	Relocate trench. Do not excavate near watercourse.
Reduction Design to reduce identified hazards but beware of increasing others	Install underpinning or temporary works ahead of trenching.	Redirect watercourse, reinforce banks, extend flood defences.
Control Design to provide acceptable safeguards for all remaining identified hazards	Monitor movement and specify that appropriate plant and materials always to be available.	Allow for the early construction of an appropriate system of drains and pumps.

Examples of risk mitigation (issues addressed at different stages)

A communications company wants to provide the option of cable TV to all properties in a township of historical interest.

Concept design	Scheme design	Detail design
Can the existing telephone cable network be used? Is there major work planned by other communication/utility companies?	Consider trench sharing with other services. Keep away from existing services and use combined dig-and-lay techniques.	Programme the works to avoid bad weather, local events, and tourist seasons.

A two-storey townhouse is to be constructed in an urban setting. Properties are located on each side and to the rear. The site has a road frontage. Excavation for strip footing foundations is in topsoil/clay.

Concept design	Scheme design	Detail design
Layout to consider ground slope, services location, access. Consider also adjacent buildings/structures.	Consider stepped footings on sloping ground. Services, water table, stockpile of excavated material.	Stepped footings reduce excavated depth. Drainage measures to reduce flooding risk. Plant location during concreting. Use concrete backfill to reduce settlement.

B3 Group B – Excavations and foundations
3 – Trenches for foundations and services

Related issues

References within this document

References	Related issues
A1	Local environment should be protected from the hazards associated with excavations. Identify also location of services/building utilities.
A3	Ground investigations should include testing for toxic materials, soil density, gases, stability and ground water details.
A4	Access/egress routes should be kept clean, clear of obstructions and be suitable for emergency services.
A5	Site layout should be related to the working space available, designer's assumptions of excavation plant to be used and the size of the site.
E1	Working space on small sites accentuates the hazards of excavation. Consider practicalities of excavation and construction sequence.
E2,E3	Excavation adjacent to roads and railways should include safe clearance distances and/or suitable protection for site personnel.
E6,E7	Working near water introduces a greater risk that the excavation may flood.

References for further guidance

Primary general references and background information given in Section 4

Primary

- CIRIA R97 (1993) Trenching practice. Second edition
- CIRIA SP147 (1998) Trenchless and minimum excavation techniques: planning and selection
- HSE CIS8 (revised) Safety in excavations
- DOT Safety at street works and road works: A code of practice, 2001

Secondary

- HSE HSG47 Avoiding danger from underground services
- HSE HSG66 Protection of workers and the general public during the development of contaminated land
- HSE INDG258 Safe work in confined spaces
- HSE GS6 Avoidance of danger from overhead electric power lines

Background

- CIRIA R130 (1993) Methane: its occurrence and hazards in construction
- CIRIA TN112 (1983) Pipe jacking: a state-of-the-art review (microfiche)

Classification

CAWS Group D,P&R **CEWS** Classes E,I,J,K,L&Z **CI/SfB** Code (1-)C

Group B – Excavations and foundations
4 – Retaining walls

B4

Read the Introduction before using the following guidelines

Scope

- Single-storey basements, drainage chambers, service ducts.
- Earthworks/cuttings: roadworks, railways, watercourses, landscaping.
- Temporary retaining walls, eg sheet piles.

Exclusions

- Deep basements and shafts (see B2).
- Cofferdams and caissons (see E7).
- Mining and work adjacent to deep water.
- Gabions.

Major hazards

Refer to the Introduction for details of accident types and health risks

Buried/crushed/trapped
Collapse of unretained material.
Collapse of incomplete and temporarily unstable wall.
Inadequate working space behind wall.

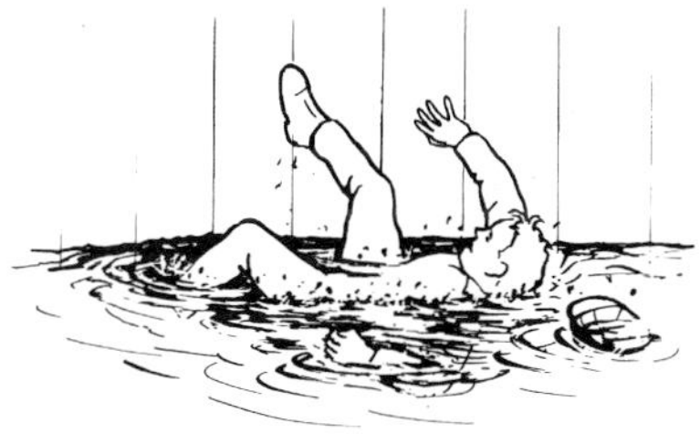

Falls
Into base excavation from top of incomplete wall.
Into water-filled excavation or temporarily-retained water.

Plant and transport
Site traffic. Slewing of concrete skip, shutters, etc.
Work adjacent to road/railway.

Services
Electrocution.
Gas pipelines.
Sewage and process effluents.
Emergency. communications.

Ground movements
Subsidence of adjacent structures.
Fractured services and waste pipes.

Health hazards
Noise/vibration.
Dust inhalation.
Fumes.
Water-borne diseases.
Manual handling.

Group B – Excavations and foundations
4 – Retaining walls

Specific hazard identification

Possible key considerations:

- **Site survey** - Where are service? What structures are adjacent to site? What is topography?
- **Ground investigation** - Is there evidence of earth movement? Where is water table?
- **Environment** - What are implications of site location? (eg urban, rural, next to river, railway, road).
- **Retaining wall** - What is wall to be used for? What type is it to be (eg gravity, reinforced concrete, temporary)?
- **Plant & Machinery** - What is sufficient working space for the proposed construction methods?
- **Sequences** - What are the construction sequences?
- **Access** - What access is required to rear of wall?
- **Any others?**

Prompts

- Soil types
- Services
- Adjacent structures (foundations)
- Retaining wall type
- Plant/machinery
- Access/working space
- Weather
- Backfilling

Hazards consideration in design

<table>
<tr><th>Stage</th><th>Considerations and issues</th><th>Possible design options to avoid or mitigate hazards identified for retaining walls</th></tr>
<tr><td>Concept design</td><td rowspan="3">• Why retaining wall?
• Location
• Environment
• Topography
• Access
• Services
• Wall position
• Protection of property
• Adjacent structures
• Working space
• Temporary roads
• Ground investigation
• Retaining wall types
• Plant/machinery
• Backfilling
• Drainage
• Ground stability
• Maintenance
• Advance works
• Temporary live loads during construction</td><td>• Position could determine wall type that can be practically and safely constructed: examine options available.
• Consider type of wall to minimise hazard of interaction with services, particularly gas and electricity.
• Evaluate possible effects on adjacent structures during operation.</td></tr>
<tr><td>Scheme design</td><td>• Design to minimise depth of excavations (eg incorporate a piled base or grout curtain).
• Consider excavation techniques in relation to ground conditions, working space and environmental constraints. (Relate to wall design.)
• Identify necessary ground and surface water control systems.
• Eliminate need for operatives to access rear of wall in confined spaces.
• Backfilling prohibited or limited until a certain stage of wall construction.</td></tr>
<tr><td>Detail design</td><td>• Consider temporary diversion of services during retaining wall construction.
• Wall design to include facilities for ease of maintenance.
• Prefabricated units to provide fixing locations for permanent or temporary guard rails.</td></tr>
</table>

Examples of risk mitigation (methods of solution)

ACTION / ISSUE	Ground slippage	Gas pipes crossing line of wall
Avoidance Design to avoid identified hazards but beware of introducing others	Pre-stabilise ground.	Change route of wall to avoid gas pipes.
Reduction Design to reduce identified hazards but beware of increasing others	Consider piles, soil nailing, ground anchors.	Change design of wall in vicinity of gas pipes (eg change ground levels or select shallower and wider base).
Control Design to provide acceptable safeguards for all remaining identified hazards	Outline possible staged construction sequence.	Specify robust protection to be incorporated into the permanent works.

Examples of risk mitigation (issues addressed at different stages)

An existing busy urban road is to be widened. Land is not available either side. The widening is therefore to be completed on line. The existing cutting slope will require a retaining wall (height 5 metres).

Concept design	Scheme design	Detail design
Consider environmentally appropriate solutions and proposals for existing road traffic. Site history - slope stability.	Bored piles reduce excavation hazard and noise and effects on existing traffic; locate services.	Access/egress site details. Allow for temporary diversion of traffic/ lane closures.

A building is to be constructed within an urban environment. Car parking is to be made available in the basement. The surrounding properties are shops, offices, residential blocks.

Concept design	Scheme design	Detail design
Consider proximity of adjacent properties and road traffic. Relate to environment and location. Noise and vibration issues.	Consider constructing a retaining wall prior to basement excavation. Ground investigation (water-bearing strata). Consider diaphragm wall, combined with ground anchors.	Precast panels/in-situ concrete (underpin adjacent building).

Group B – Excavations and foundations
4 – Retaining walls

Related issues

References within this document

References	Related issues
A1	Local environment should be protected from the hazards associated with excavations.
A3	Ground investigation should include tests for toxic materials, soil density, gases, stability and groundwater details.
A4	Access/egress routes should be kept clean, clear of obstructions and be suitable for emergency services.
A5	Site layout should be related to the working space available, the excavation plant to be used and the size of the site.
E1	Working space on small sites accentuates the hazards of excavation.
E2,E3	Excavation adjacent to roads and railways should include safe clearance distances and/or suitable protection for site personnel.
E6,E7	Working near water introduces a greater risk that the excavation may flood.
B5	Ground stabilisation may provide a safer solution.
	Underground/overhead services need to be carefully considered.

References for further guidance

Primary general references and background information given in Section 4

Primary

- HSE INDG258 Safe work in confined spaces
- HSE GS6 Avoidance of danger from overhead power lines
- HSE GS29 Parts 1 to 4 Health and safety in demolition work
- HSE CIS8 (revised) Safety in excavations
- DOT Safety at street works and road works: A code of practice, 2001
- CIRIA C580 (2003) Embedded retaining walls – guidance for economic design

Secondary

- HSE HSG47 Avoiding danger from underground services
- HSE HSG66 Protection of workers and the general public during the development of contaminated land

Background

- BS 8002:1994 Earth Retaining Structures
- BS 8006:1995 Code of Practice for strengthened/reinforced soils and other fills

Classification

CAWS Group E, **CEWS** Classes E,F&H, **CI/SfB** Code (16)

Group B – Excavations and foundations
5 – Ground stabilisation

Read the Introduction before using the following guidelines

Scope

- Vibroflotation (vibrocompaction, vibroreplacement).
- Dynamic compaction.
- Lime and cement stabilisation.
- Dewatering/band drains.
- Preloading.
- Jet grouting, permeation grouting (to improve bearing capacity).

Exclusions

- Ground freezing.
- Stabilisation of rock masses (rock bolting, etc.).
- Slope stabilisation (soil nailing, retaining structures, etc.).
- Settlement reducing/elimination measures (eg compensation grouting).

Major hazards

Refer to the Introduction for details of accident types and health risks

Plant instability
Inadequate working platform - consider bearing capacity, gradient, stability of fill (settlement), slope stability.

Crushed/trapped
Manoeuvring and slewing of heavy plant and equipment.

Working environment
Noise, vibration, fumes, asphyxiation (eg nitrogen), high pressure lines and hoses.

Services
Buried underground and overhead services including gas, electricity, water. Damage by drilling, driving, impact, vibration, settlement.

Health hazards
Contact with contaminated arisings/groundwater.
Dust and irritation from materials (eg cement and lime).

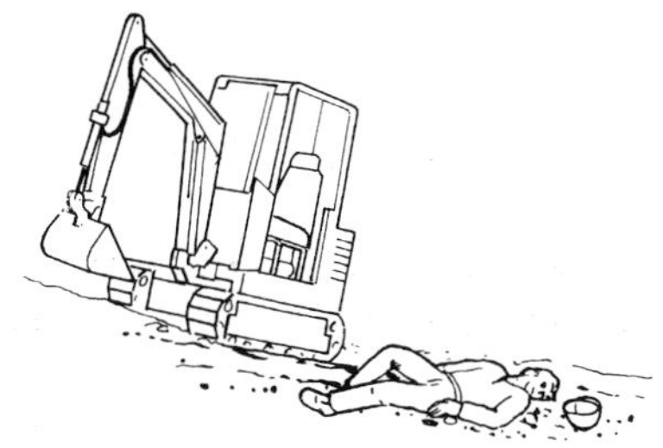

Falls
From plant and machinery tilted by uneven/soft ground.

Group B – Excavations and foundations
5 – Ground stabilisation

Specific hazard identification

Possible key considerations:

- **Access** - What provisions are necessary at each location?
- **Working space** - What needs to be provided?
- **Working platform** - Where and what is needed?
- **Services** - Where are they located?
- **Stability** - What considerations need to be made for nearby structures?
- **Workers/public** - What is their proximity?
- **Workers** - What is proximity of workers during operations?
- **Contamination** - Where is affected ground and water?
- **Materials** - What substances are hazardous to health?
- **Any others?**

Prompts

- Access routes
- Plant size and weight
- Site survey rig details
- Services check
- Structural survey
- Noise/dust/lifting materials
- Site investigation

Hazards consideration in design

Stage	Considerations and issues	Possible design options to avoid or mitigate hazards identified for ground stabilisation
Concept design	• Why stabilise? • Location of structure • Overall programme • Type of structure • Surveys/site investigation/desk study • Access • Programme/sequences • Interfaces • Specific alternatives • Depth of stabilisation • Type of stabilisation material • Access/haul routes • Working platform	• Review reasons for stabilising (eg bearing capacity or settlement). • Identify constraints and consider relocation of structure. • Consider alternative types of structure, such as flexible foundations. • Surveys/site investigation/desk study to enable options to be assessed systematically.
Scheme design		• Consider access requirements and location relative to public. • Eliminate interface constraints. • Consider alternative stabilisation methods? (Contact experienced specialist contractors for advice.)
Detail design		• Consider depths of treatment required by design, in relation to plant capabilities (eg can smaller plant be used?) • Can an alternative stabilisation material be used (eg cement instead of lime)? • Can same access points and haul roads be used throughout contract? • Can working platform form part of permanent works?

Examples of risk mitigation (methods of solution)

ACTION \ ISSUE	Use of surcharging	Use of lime piles
Avoidance Design to avoid identified hazards but beware of introducing others	Avoid stabilisation by phased construction (eg allow periods for the ground to consolidate between defined stages of construction).	Avoid hazard of lime piles by using stone columns?
Reduction Design to reduce identified hazards but beware of increasing others	Consider preloading or drainage to induce settlement.	Minimise depth of ground treatment to allow lighter/smaller plant to be used?
Control Design to provide acceptable safeguards for all remaining identified hazards	Outline appropriate monitoring locations and criteria. Change the building/structure performance specification for movement.	Outline hazards of plant instability and need to design working platform.

Examples of risk mitigation (issues addressed at different stages)

A haul road at a supermarket site is to be constructed on a wet clay subgrade, to enable use of road lorries/tippers. To save on granular material, it is proposed to stabilise the clay subgrade using quick-lime.

Concept design	Scheme design	Detail design
Consider re-aligning haul road and the use of all terrain dump trucks.	To avoid hazard of anhydrous quick-lime, consider alternative forms of stabilisation, eg a) Geogrid reinforced, granular roadway with geotextile separation membrane. b) Re-sequence earthmoving to summer period.	To reduce risk associated with exposure to quick-lime, specify lime graded between 3mm and 12mm (to reduce dusting).

An embankment adjacent to an estuary is to be founded on soft clay. It is proposed that vibro-replacement with stone columns will be used to reduce settlement of the embankment.

Concept design	Scheme design	Detail design
To avoid hazard of plant instability, consider re-locating embankment onto stronger ground or the use of lightweight fill to reduce loading.	To avoid hazard of heavy plant (70 tonne) operating on soft material, consider differential and total settlement criteria to see if a shallow foundation type can be adopted (eg. geocell mattress/raft).	Consider using smaller plant. Outline need to plan/sequence work carefully (eg work along embankment using stone columns to act as foundation to haul road and working platform for vibroreplacement plant). Separate public and workers from haul road (reroute footpaths)

Group B – Excavations and foundations
5 – Ground stabilisation

A GENERAL PLANNING
B 5 EXCAVATIONS AND FOUNDATIONS
C PRIMARY STRUCTURE
D BUILDING ELEMENTS AND BUILDING SERVICES
E CIVIL ENGINEERING

Related issues

References within this document

References	Related issues
A1	Surrounding environment will be affected by ground stabilisation work.
A2	Site clearance works may be able to accommodate access provision for future stabilisation work.
A3	Site investigation will be necessary to determine the criteria for future stabilisation work.
A4	Access onto the site will be necessary for the plant.
A5	Site layout determines the extent of the stabilisation necessary.
B1	General excavation may require the use of stabilisation measures.
E2	Working near roads may be undertaken while road is in use.
E3	Working near railways may be undertaken while railway is in use.
E6,E7	Working near water may be a major consideration.
	Contamination may be caused by ground treatment processes (eg groundwater).

References for further guidance

Primary general references and background information given in Section 4

Primary

- CIRIA SP123 (1996) Soil reinforcement with geotextiles
- CIRIA B11 (1991) Prefabricated vertical drains
- CIRIA R95 (1982) Health and safety aspects of ground treatment materials (microfiche)
- ICE Works Construction Guide "Ground stabilisation, deep compaction and grouting"

Secondary

- CIRIA R113 (1986) Control of groundwater for temporary works (microfiche)
- CIRIA C514 (2000) Grouting for ground engineering
- HSE CIS8 (revised) Safety in excavations

Background

- Construction Industry Training Board CITB/FPS Document GE 708 Safety on piling sites
- CIRIA SP136 (1996) Site guide to construction of foundations
- Bell, F G (1993) Engineering treatment of soils (Spon)
- Moseley, M P, Ground improvement (Blackie)

Classification

CAWS Group D, **CEWS** Classes E&C, **CI/SfB** Code (11)

Group B – Excavations and foundations
6 – Piling

Read the Introduction before using the following guidelines

Scope

- Foundation piles - bored, driven, timber, steel, concrete.
- Load testing of piles.
- Shafts < 3m diameter.
- Installation of sheet piles.

Exclusions

- Piles for underpinning (see B7).
- Piling over water (see E6).
- Piling next to railways and/or roads (see E2, E3).

Major hazards

Refer to the Introduction for details of accident types and health risks

Plant instability
Inadequate working platform for equipment - considerations such as: bearing capacity, gradient, stability, variability of ground conditions.

Plant and Machinery
Lifting, slewing and pitching of casings, piles and reinforcement cages. Movement of piling rigs, Delivery of materials.

Working environment
Noise, vibration, exhaust fumes, trips, open bores/excavations, impact from spoil falling off auger.

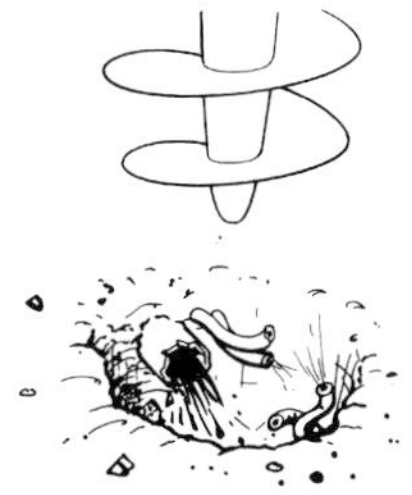

Services
Buried underground and overhead services including gas, electricity, water and drainage by drilling, driving, impact, vibration, settlement.

Health hazards
Contact with contaminated arisings/groundwater.
Dust and irritation from materials (eg bentonite and cement).

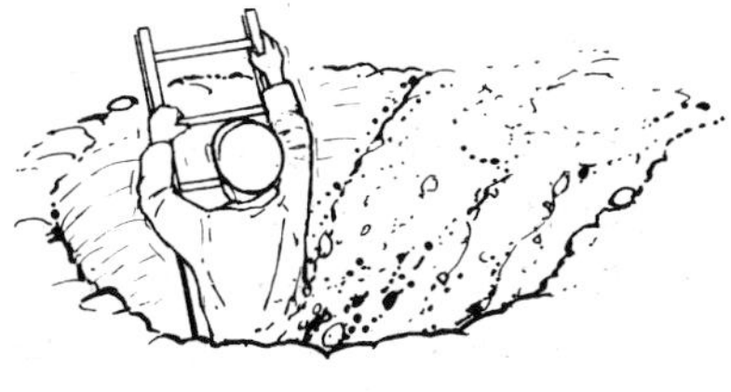

Collapse of excavation
Insufficient support of bores/excavation results in collapse, including loss of slimy support materials (eg drains). Collapse during inspection of pile base, or collapse due to construction delays.

B6 Group B – Excavations and foundations
6 – Piling

Specific hazard identification

Possible key considerations:

- **Access** - What provision is required at each pile location?
- **Working space** - What needs to be provided?
- **Working platform** - Where and what is needed?
- **Services** - Where are they located?
- **Stability** - What considerations need to be made for nearby structure?
- **Workers/public** - What is the proximity?
- **Contamination** - Where is affected ground and water?
- **Materials** - What substances are hazardous to health?
- **Open pile borings** - How will these be protected?
- **Operations** - What sequences will be adopted?
- **Obstructions** - Check for underground structures and for abandoned foundations, works and services.
- **Any others?**

Prompts

- Access routes
- Plant size
- Site survey/rig details
- Services check
- Structural survey
- Noise/dust
- Lifting materials/crane loads
- Site investigation
- Confined space/falling materials

Hazards consideration in design

Stage	Considerations and issues	Possible design options to avoid or mitigate hazards identified for piling
Concept design	• Why is piling required? • Location of structure • Type of structure • Overall programme • Surveys/site investigation/desk study • Access • Programme/sequences • Interfaces • Type of pile • Working platform • Pile length • Obstructions • Programme • Temporary works	• Review the reasons for piling (eg bearing capacity or settlement?). • Identify constraints and consider relocation of structure. • Consider alternatives (eg shallow foundations, rafts or flexible foundations such as a reinforced soil mattress). • Surveys/site investigation/desk study to enable different options to be assessed systematically).
Scheme design		• Consider piling techniques to minimise noise and spreading of spoil (if ground contaminated). • Identify access requirements associated with equipment (eg overhead power lines and low headroom). • Programme to allow separation of site activities such as trial piling and demolition (programme and interface constraints). • Consider the magnitude and distribution of structural loads in relation to constraints (eg low headroom may restrict pile size).
Detail design		• Consider support of bore. • How will obstructions such as old basements or unforeseen objects be removed if encountered? • Allow sufficient programme time for method (eg Continuous Flight Auger (CFA) bored piles are quick to construct).

Examples of risk mitigation (methods of solution)

ACTION \ ISSUE	Piling plant on soft ground	Piling near overhead power lines
Avoidance Design to avoid identified hazards but beware of introducing others	Consider alternative foundation to avoid hazard of piling plant.	Avoid piling near power lines.
Reduction Design to reduce identified hazards but beware of increasing others	Use transfer structure to reduce number of piles, and avoid piling in the softest area.	Design alternative forms of foundations.
Control Design to provide acceptable safeguards for all remaining identified hazards	Outline allowable bearing capacity of ground.	Highlight the hazards.

Examples of risk mitigation (issues addressed at different stages)

An embankment is to be located beneath overhead electricity lines and adjacent to pylons. It is proposed to use bored cast in-situ concrete piles.

Concept design	Scheme design	Detail design
To avoid hazard of electrocution, consider moving structure or diverting overhead power lines. If this is not possible reconsider design criteria such as differential settlement rather than total settlement.	Design developed using an increased number of smaller piles with a transfer structure. Adopt bored piles rather than driven piles to avoid hazard of pitching piles.	Use tripod rigs/cut down mast on piling rig to enable safety zone to be maintained. Detail pile reinforcement to suit headroom restrictions.

Building layout requires high column loads to be supported. It is proposed to adopt large diameter under-reamed bored piles. Consider hazard of pile bore instability during base inspection.

Concept design	Scheme design	Detail design
Consider a design that uses transfer structures to allow use of more, smaller diameter, straight shafted piles. Carry out trial pile to check underream stability?	Consider use of larger diameter shaft and smaller diameter underream. Consider reducing number of underream piles by having fewer larger diameter piles. Down rate pile base and accept inspection by CCTV. Use base grouting of pile?	Design casing to support pile shaft. Make inspection cage equal in height to uncased height. Specification of unavoidable inspection requirements to favour sampling and testing of the shortest duration possible.

B6 Group B – Excavations and foundations
6 – Piling

Related issues

References within this document

References	Related issues
A1	Surrounding environment will be affected by piling operations.
A2	Site clearance works should accommodate access provisions for future piling works.
A3	Site investigation will be necessary to determine data for piling work.
A2,A3	Ground contamination will influence choice of piling.
A4	Access requirements for piling equipment is a major consideration.
A5	Site layout may influence piling method.
B1	General excavation requirements may influence piling methods and equipment.
C1,C2	Concrete methods need to be considered.
E2	Working near roads may have to be undertaken while road is in use.
E3	Working near railways is usually undertaken while railway is in use.
E6,E7	Working near water may be a major consideration.
	Underground services are a major factor in piling operations.

References for further guidance

Primary general references and background information given in Section 4

Primary

- Construction Industry Training Board CITB/FPS Document GE 708 Safety on piling sites
- BS 8008:1996 A guide to safety precautions and procedures for the construction and descent of machine bored shafts for piling and other purposes
- CIRIA PG1 (1988) A review of bearing pile types
- CIRIA C703 (2003) Crane stability on site. Second edition

Secondary

- CIRIA PG2–9 Piling guides (some microfiche only)
- HSE CIS8 (revised) Safety in excavations

Background

- CIRIA SP105 (1995) Remedial treatment for contaminated land, Vol V: Excavation and disposal
- Fleming, W G K, Piling engineering (Blackie)
- Tomlinson, M J, Pile design and construction (Spon)

Classification

CAWS Group D, **CEWS** Classes O,P,N&Q, **CI/SfB** Code (17)

Group B – Excavations and foundations
7 – Underpinning

B7

Read the Introduction before using the following guidelines

Scope

- Mass concrete underpinning.
- Pier and beam underpinning.
- Pile and beam underpinning.
- Mini-piling.

Exclusions

- Jet grouting and permeation grouting (see B5).

Major hazards

Refer to the Introduction for details of accident types and health risks

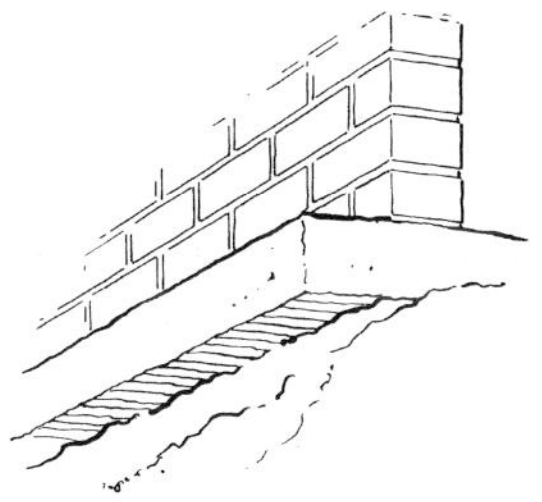

Structural instability
Inadequate stability of structure and risk to workers and occupants of the buildings.

Crushed/trapped
Narrow spaces and low headroom for manoeuvring heavy plant and materials. Manual handling of casings, reinforcement, trench sheets and concrete in confined spaces.

Working environment
Noise, vibration, fumes (confined space ventilation?).
Use of tools in confined spaces.

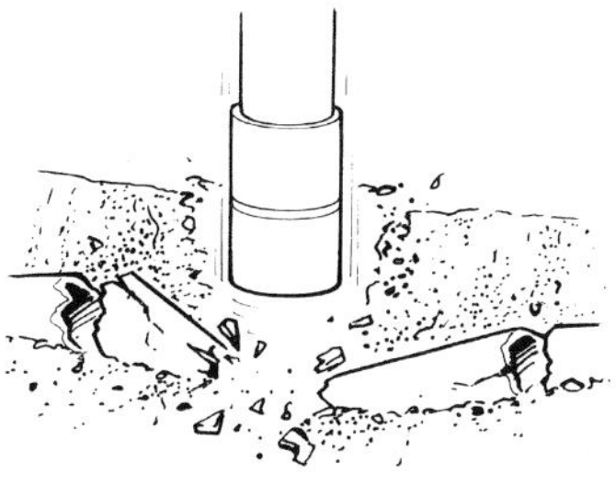

Services
Buried underground services and overhead services including gas, electricity, water inside and outside buildings. Damage by drilling, driving, impact, vibration, settlement.

Health hazards
Contact with contaminated arisings/groundwater, cement, concrete, adhesives, including inhalation of fumes and dusts (eg fibres from dry materials.

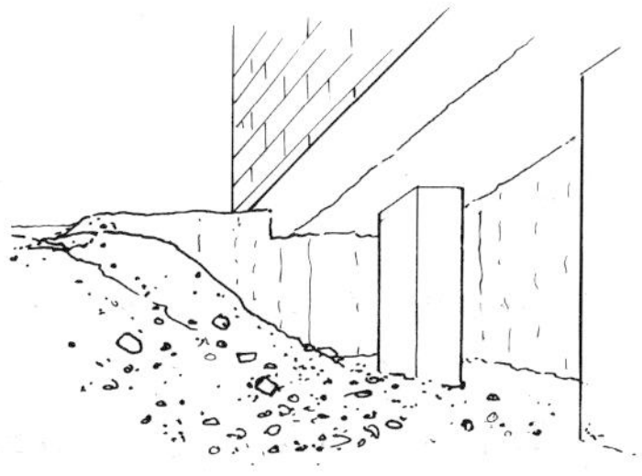

Collapse of excavations
Insufficient support results in collapse with risk of entrapment or falls and sudden water inflow/flooding.

Group B – Excavations and foundations
7 – Underpinning

Specific hazard identification

Possible key considerations:

- **Access** - What provisions are necessary?
- **Working space** - What needs to be provided?
- **Services** - Where are they located?
- **Stability** - What is the condition of brickwork/walls?
- **Workers/public** - What is the proximity?
- **Contamination** - Where is affected ground and water?
- **Materials** - Where are substances hazardous to health?
- **Manual handling** - What is the extent of this aspect?
- **Hand tools** - What provisions need to be made?
- **Any others?**

Prompts

- Access routes
- Plant size
- Services
- Structural survey
- Noise/dust
- Handling materials
- Site investigation

Hazards consideration in design

Stage	Considerations and issues	Possible design options to avoid or mitigate hazards identified for underpinning
Concept design	• Why underpin? • Type of structure • Access • Interfaces • Surveys/site investigation/desk study • Type of underpinning • Stability during underpinning • Sequences • Handling materials • Work sequences • Detailing • Temporary works • Monitoring	• Consider strengthening and/or repair of structure instead of underpinning. • Consider removing cause of distress and repairing structure. • Consider stabilising ground (eg grouting old mineworkings). • Structural survey/site investigation/desk study to understand problem systematically.
Scheme design		• Choice of shallow foundations may reduce hazard of entrapment. • Choice of mini-piling may reduce excavation of contaminated material. • Excavation may destabilise structure. • Consider mini-piles and needling, which may cause less disturbance to a fragile structure.
Detail design		• Define sequence of support to avoid instability. • Avoid starter bars. • Require monitoring of structure/ground during work, having defined allowable movements.

Examples of risk mitigation (methods of solution)

ACTION \ ISSUE	Need to underpin	Mass concrete underpinning
Avoidance Design to avoid identified hazards but beware of introducing others	Avoid underpinning by addressing cause of ground movement and repairing structure.	Avoid entrapment by avoiding descent of excavations (remote operation and inspection).
Reduction Design to reduce identified hazards but beware of increasing others	Consider different types of underpinning available and hazards with each.	Reduce hazard of excavation by minimising descent.
Control Design to provide acceptable safeguards for all remaining identified hazards	Outline interface issues (time, space etc). Associated with types of underpinning.	Outline need to control groundwater inflow in H&S Plan.

Examples of risk mitigation (issues addressed at different stages)

Structure constructed on varying depth of made ground is to be underpinned.

Concept design	Scheme design	Detail design
Mass concrete underpinning chosen to avoid hazard of piling plant. Underpinning to be continuous, carried out in a "hit-and-miss" manner.	Consider dowels protruding into "miss" bays/ excavation. Design for use of: - bar couplers/reinforcement coupler strips, or - "joggle/key" joint - smaller diameter mild steel dowel bars (10/12mm diameter) that can be bent out of way.	Choose small diameter dowels. Control hazard by placing "mushroom" caps on projecting dowel bars to protect workers in trenches.

Structure on contaminated ground is to be underpinned.

Concept design	Scheme design	Detail design
The design is to consider techniques that do not involve excavation. Can the structure be strengthened or repaired without excavation?	The design will include an underpinning technique that minimises excavation and contact with contaminated material (eg mini-piles?). Identify contaminants by site investigation.	Highlight need to ventilate confined spaces inside building during work. Identify need to monitor confined spaces.

Related issues

References within this document

References	Related issues
A1	Surrounding environment may be affected by the underpinning work.
A2	Site clearance may be moved before work can commence.
A3	Site investigation will be necessary to identify scope of work.
A4	Access onto the site will be a major consideration.
A5	Site layout may influence the underpinning method and temporary works.
B1	General excavation may be necessary.
B6	Piling may be used as a method of support.
C1,C2	Concrete pumping and use will need to be considered.
	Groundwater level and surcharge from plant will influence stability of trench excavations.
	Contaminated ground will influence design and precautions.
E3	Working near railways will be a major consideration in the choice of plant..
E6,E7	Working near water will involve possibility of dewatering excavation.

References for further guidance

Primary general references and background information given in Section 4

Primary

- CITB Construction Site Safety Note 10 - Excavations
- Thornburn S, and Littlejohn G S, Underpinning and retention (Blackie)
- HSE CIS8 (revised) Safety in excavations
- CIRIA R97 (1992) Trenching practice

Secondary

- BRE Digest 381 Site investigation of low rise buildings – trial pits

Background

- CIRIA TN95 (1986) Proprietary trench support systems (microfiche only)
- Thorburn, S and Littlejohn, G S (1993) Underpinning and retention (Blackie)
- Bullivant, R and Bradbury, N W (1996) Underpinning: a practical guide (Blackwell)

Classification

CAWS Group D, **CEWS** Class C,F&P, **CI/SfB** Code (1-)B

Group C – Primary structure
1 – General concrete

C1

Read the Introduction before using the following guidelines

Scope

- Construction/maintenance/demolition procedures.
- Consideration of working at height or depth.
- Size of any pour, construction joints, expansion joints, movement joints and their positions.
- Element size/length/weight.
- Ducts, holes, chases, cutting, drilling.
- Falsework, formwork, shuttering.
- Bearings/connections/movement joints, water bars.
- Surface preparation, grit blasting, chemical retarders, expanded metal.
- Prefabrication – advantage, reduction of different activities.

Exclusions

- Ornamental concrete.
- Gunite/shotcrete.
- Grout.

Major hazards

Refer to the Introduction for details of accident types and health risks

Falling from height
Operatives falling during: preparation of falsework, formwork, concreting, curing.

Health hazards
Cement dust.
Wet concrete:
- skin/eye irritation,
- caustic burning.

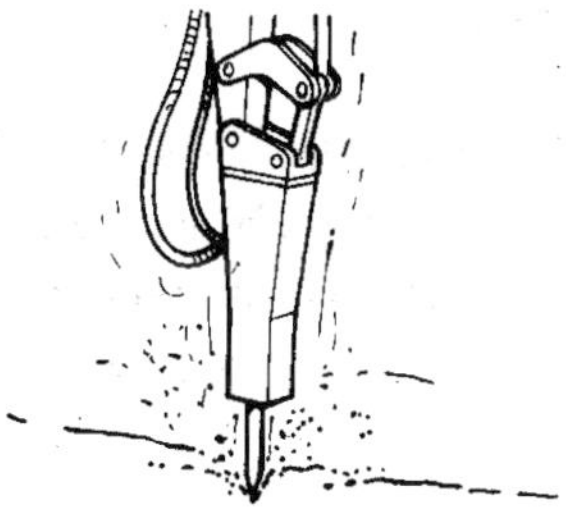

Noise and vibration
Concrete:
- scabbling,
- breaking.

Inhalation of dust
Concrete :
- demolition,
- drilling/cutting,
- scabbling.

Impact
Concrete vibrators.
Concrete pumps.
Moving plant contact.
Skips, dumpers.

Fumes/heat
Cutting/thermic lances:
- inhalation,
- burns.

Group C – Primary structure
1 – General concrete

Specific hazard identification

Possible key considerations:

- **Materials choice** – Show that concrete is the appropriate material.
- **Properties** – Can elements, properties, features be standardised to simplify working procedures?
- **Prefabrication** – Show off-site fabrication design has been maximised.
- **Procedure** – Show that material is being handled appropriately.
- **Materials** – When is the concrete deleterious to: skin, eyes, inhalation, ingestion?
- **Climate** – Show that effects will not be exacerbated by sun, wind, cold, heat or water.
- **Emissions** – Will the process cause: fire, fumes, spray, dust, toxins?
- **Protection** – Has the need for additional protection been identified: enclosure, extraction, isolation, personal protection?
- **Any others?**

Prompts

- Location
- At height/depth
- Materials properties
- Element size
- Prefabrication
- Construction joints
- Holes/services
- Cutting
- Fixings
- Weight
- Demolition
- Access

Hazards consideration in design

Stage	Considerations and issues	Possible design options to avoid or mitigate hazards identified for general concrete
Concept Design	• Why use concrete? • Scope of use • Familiar? - materials - application - location • Environment • Materials • Demand on resources	• Having chosen to use concrete, will it be identified where it will be mixed and placed? Can it be safely located in that position? • If in-situ concrete, does the site have sufficient space for batching or is ready – mixed concrete available, within reasonable access to the point of placing? • Is the way in which the material is to be used known to the proposed contractors? • Will hazardous processes be required? eg hydrocutting, thermic lance
Scheme Design	• Technology known? • Characteristics • Complexity • Programme • Review • Process • Protection • In-situ concrete • Pre-cast concrete • Pre-stressed concrete	• Are special concrete mixes being used and can they be supplied locally to the site? • Consider the programme: where and when will the concreting be carried out? - outside or undercover? - in summer or winter? • Will concreting be at ground level and units raised or lowered? or prepared and concreted at height or depth? • Do sequences and finishes require scabbling?
Detail Design	• Elements • Standardisation • Features • Holes/fixing • Cutting • Constituents • Method • Plan/control • Testing • Monitoring • Analysing	• If the design requires a specific method of build, set it down for the contractor. • A safe method of construction should be determined by the designer. • Optimise standard unit details to minimise errors. • Include details for relevant stages in the life of the concrete element, construction, use, maintenance and demolition, chases, fixings, conduits, etc. • Chases or holes should be detailed to be cast into the unit and not cut or drilled later.

Examples of risk mitigation (methods of solution)

ACTION \ ISSUE	Concrete frame building	Working at height to install embedded fixing in RC element
Avoidance Design to avoid identified hazards but beware of introducing others	Avoid working at height by fabricating elements off-site.	Fixing installed in element off-site to avoid working at height.
Reduction Design to reduce identified hazards but beware of increasing others	Element prefabricated (eg reinforcement). Fabricate at ground level and lift.	(Does not apply if risk can be avoided)
Control Design to provide acceptable safeguards for all remaining identified hazards	Provide information of element construction.	(Does not apply if risk can be avoided)

Examples of risk mitigation (issues addressed at different stages)

An existing bridge over a busy road is to be upgraded. The abutments are adequate but the concrete deck and beams need to be replaced. The major risk is working at height and restricted access.

Concept design	Scheme design	Detail design
Consider fabrication off-site to avoid site activities at height over the road.	The beams will be pre-cast concrete. Tee beams chosen to provide immediate access for operatives to the bridgedeck.	Pre-cast units to be designed with service ducts, street furniture fixings and temporary protection included, to avoid on-site activities.

Concrete chambers are required to facilitate maintenance inspection of underground services. The variable ground conditions and limited working space indicate that working at depth within the excavations is a major risk.

Concept design	Scheme design	Detail design
Hazard studies have indicated that inspection chambers should be reinforced concrete to reduce hazardous activities in excavations, at depth.	To reduce hazards of depths, the inspection chambers will be pre-cast on-site and lowered into position. The additional hazards introduced by this method have been considered preferable to those being avoided.	Inspection pit design will include cast-in ducts, step irons, fixings, lifting points.

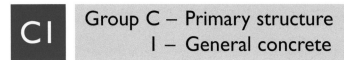

Group C – Primary structure
1 – General concrete

Related issues

References within this document

References	Related issues
A1	Surrounding environment for precautions to control spread of concrete materials.
A2	Site clearance will need controls to prevent pollution from concrete dust.
A3	Site investigation may be required to determine the extent and condition of the concrete.
A4	Access onto the site will be carefully planned to transport concrete or the constituent materials.
A5	Site layout will need to be co-ordinated with the requirement for concreting facilities.
C2	Insitu concrete will need to consider this work sector.
C3	Precast concrete will consider this work sector.
C4	Prestressed concrete will consider this work sector.
D1	External cladding may be made of concrete.
D5	Surface coatings and finishes may be an application to concrete products.
D10	Vertical transportation needs to be carefully considered when transporting concrete.
E1	Civil engineering small works will most likely use concrete.
E2	Roads, working adjacent to, maintenance.
E3	Railways, working adjacent to, maintenance.
E4	Bridge construction will probably use concrete.
E5	Bridge maintenance will cover concrete elements.
E6	Working over/near water may place concrete in an aggressive environment.
E7	Cofferdams in water are likely to be concrete.

References for further guidance

Primary general references and background information given in Section 4

Primary

- CIRIA R146 (1995) Design and construction of joints in concrete structures
- HSE HSG150 Health and safety in construction
- HSE CIS26 (revised) Cement
- BS 5975:1996 Code of practice for falsework

Secondary

- CIRIA PR70 (1999) How much noise do you make? A guide to assessing and managing noise on construction sites
- CIRIA R67 (1991) Tables of minimum striking times for soffit and vertical formwork
- CIRIA R136 (1995) Formwork striking times
- CIRIA SP57 (1988) Handling of materials on site
- HSE GS28 Parts 1 to 4 Safe erection of structures

Background

- CIRIA C512 (2000) Environmental handbook for building and civil engineering projects Part 1: design and specification
- CIRIA R120 (1990) A guide to reducing the exposure of construction workers to noise
- British Cement Association (1993) Concrete on site (BCA)

Classification

CAWS Group E, **CEWS** Class F,G,H, **CI/SfB** Code (2-)q

Group C – Primary structure
2 – In-situ concrete

Read the Introduction before using the following guidelines

Scope

- Falsework – scaffolding, patent structure, steelwork, timber.
- Formwork – temporary, permanent, erecting, striking, pour sizes, construction joints.
- Reinforcement – spacing, grouping, supporting, projecting.
- Positioning concrete – skips, pumps, barrows, shute, tremmie.
- Placing concrete – spreading, levelling, loading, finishing (power float/tamp etc.).
- Concrete compaction – vibrating poker, tamper, external vibrators.
- Striking formwork/falsework – concrete maturity.
- Temporary loading conditions – temporary design, temporary support.

Exclusions

- No-fines concrete.
- Dry-lean concrete.
- Demolition.

Major hazards

Refer to the Introduction for details of accident types and health risks

Falls from height
Accessing or working on scaffold, preparing for or placing in-situ concrete, access, working on formwork.

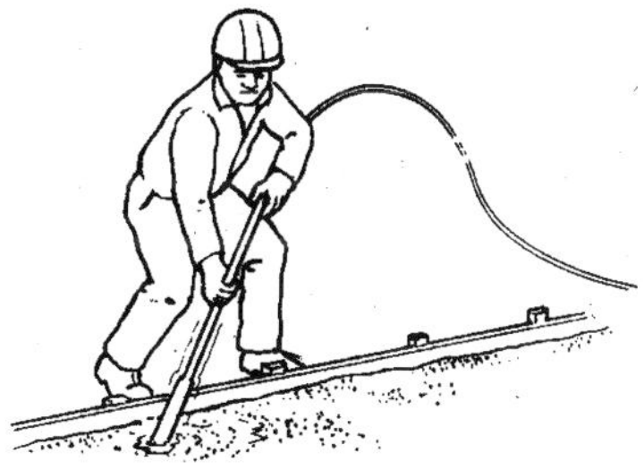

Noise, vibration
Working with breakers, vibrators, scabblers, saws, etc.

Collapse
Formwork/falsework permanent structure collapse under loading from fresh concrete, excessive spans, large sections.

Moving Plant
Collision with moving machines (concrete trucks, concrete mixers, pumps, dumpers, etc.).

Health hazards
Inhalation of dust from cement, breaking out or scabbling concrete.

Irritants
Concrete constituent materials, mould oils:
- skin/eye irritation
- caustic burns.

A GENERAL PLANNING
B EXCAVATIONS AND FOUNDATIONS
C 2 PRIMARY STRUCTURE
D BUILDING ELEMENTS AND BUILDING SERVICES
E CIVIL ENGINEERING

Specific hazard identification

Possible key considerations:

- **Conditions** – Where are all the temporary and permanent conditions listed?
- **Design** – How is it shown that the element is designed to satisfy all the temporary and permanent conditions?
- **Temporary works** – Demonstrate how hazards have been checked in falsework and formwork design.
- **Access –** Show adequate access for plant, materials and operatives has been identified.
- **Space** – Show there is adequate space for access and working.
- **Environment** – Demonstrate the working environment is satisfactory.
- **Work Access** – Where is access to work areas set out?
- **Storage –** Show that materials storage areas have been considered.
- **Prefabrication** – Show that prefabrication has been maximised.
- **Any others?**

Prompts

- Access
- Transporting materials
- Positioning materials
- Formwork/falsework
- Pour sizes
- Construction joints
- Reinforcement
- Concreting
- Compaction
- Protection
- Loading

Hazards consideration in design

Stage	Considerations and issues	Possible design options to avoid or mitigate hazards identified for insitu concrete
Concept design	• Why in-situ concrete? • Scope of work • Overall mass • Familiar to use • Accessibility set out • Interfaces defined • Characteristics	• Can the in-situ concrete work be safely constructed in the time available or should alternatives be considered (eg pre-cast concrete)? • Does the in-situ concrete work constitute a significant part of the work? If the in-situ work is not extensive, avoid sophisticated details. • Avoid using in-situ concrete at height or depth if possible.
Scheme design	• Programme plan • Temporary works • Review complexity • Span/section defined • Sequence planned • Temporary loading • Tolerances set out • Protection defined • Procedures set down	• Use a design that allows for adequate access to construct the element on site, given selected production methods. • Will the formwork for the site concreting be open and accessible, or restricted and congested? • If the design assumes a specific construction (and demolition) sequence of build, identify it clearly in the design. • Inform formwork/falsework designer of element design assumptions. • Specify sequences and finishes that do not require scabbling.
Detail design	• Falsework design • Formwork design • Access planned • Formwork detail • Falsework detail • Reinforcement • Pour sequence • Pour sizes • Curing/maturity • Striking criteria • Testing materials	• Ensure that it is possible to have safe access and place of work to construct the element. • Reinforcement cages or mats should be rigid. Avoid congestion of reinforcement in an element. • Identify reasonable pour layouts/sizes, with only reinforcement lap lengths protruding from the pour. • If the design assumes a depropping/striking time criterion (strike before/not until), ensure that it is clearly identified for the contractor in the health and safety plan. • Form chases and holes to avoid cutting or drilling later.

Examples of risk mitigation (methods of solution)

ACTION \ ISSUE	Site activities	Surface preparation
Avoidance Design to avoid identified hazards but beware of introducing others	Avoid working in confined areas.	Avoid scabbling by co-ordinating construction joints and reinforcement arrangement.
Reduction Design to reduce identified hazards but beware of increasing others	Simplify and standardise details.	Specify use of surface retarders for joint preparation.
Control Design to provide acceptable safeguards for all remaining identified hazards		

Examples of risk mitigation (issues addressed at different stages)

A framed building is to be constructed on a congested site. The frame is not a regular grid; spans and loading vary. One option is to design for a reinforced concrete frame founded on a continuous raft.

Concept design	Scheme design	Detail design
Concept reviews indicate that an in-situ concrete structure is the simplest form of construction for this irregular building, thus reducing hazards.	Design to standardise beam/slab column sizes wherever possible. Reinforcement details and construction joints planned to suit shutter assembly on site.	The specification requires the construction joint preparation to be carried out using a surface retarder to reduce noise and dust hazards from concrete surface preparation.

A reinforced concrete retaining wall is being built along the side of a road in a town centre as part of a road improvement scheme. Earlier considerations indicate that insitu concrete is the preferred option. The retaining wall has to be founded at depth in order to reach ground with sufficient bearing capacity. The design will need to take into consideration the difficult working conditions to be encountered on site in close proximity to a busy road.

Concept design	Scheme design	Detail design
The design will reduce the hazards on site by minimising and separating the number of activities required to be carried out at depth.	Reinforcement and formwork will be fabricated off-site, delivered in panels at the surface and lowered into position, to reduce activities at depth.	Control procedures will be required to monitor air quality in the excavation, which will be affected by traffic exhaust gases.

Group C – Primary structure 2 – In-situ concrete

A GENERAL PLANNING
B EXCAVATIONS AND FOUNDATIONS
C 2 PRIMARY STRUCTURE
D BUILDING ELEMENTS AND BUILDING SERVICES
E CIVIL ENGINEERING

Related issues

References within this document

References	Related issues
A1	Surrounding environment will need to be considered when examining access to the pour location.
A2	Site clearance and simple demolition may be relevant in order that in-situ concreting activities can proceed.
A3	Site investigation may have to be carried out before the criteria for the design of the reinforced concrete works can be established.
A4	Access on the site will need to be determined to ensure the safe construction of the in-situ concrete.
A5	Site layout intersects with decisions to use in-situ concrete.
B2	Deep basements shafts are likely to use in-situ concrete construction.
B4	Retaining walls are likely to use in-situ concrete construction.
B6	Piling should be considered, as bored piling will use in-situ concrete construction.
B7	Under-pinning works will use in-situ concrete construction.
C1	General concrete
C3	Pre-cast concrete may be considered as an alternative to in-situ reinforced concrete.
C4	Pre-stressed, post-tensioned concrete may be also considered as an alternative to reinforced concrete.
E1	Civil engineering, small works are likely to be using in-situ reinforced concrete structures.
E4	Bridge construction may well be constructed in in-situ reinforced concrete.
E5	Bridge maintenance will probably require periodic attention to in-situ concrete details.
E6	Working over/near water is likely to use in-situ concrete.
E7	Cofferdams in water may use in-situ concrete.

References for further guidance

Primary general references and background information given in Section 4

Primary

- HSE HSG150 Health and safety in construction
- HSE HSG32 Safety in falsework for insitu beams and slabs
- HSE CIS26 (revised) Cement
- BS 5975:1996 Code of practice for falsework

Secondary

- HSE GS28 Parts 1 to 4 Safe erection of structures
- CIRIA R67 (1991) Tables of minimum striking times for soffit and vertical formwork
- CIRIA R136 (1995) Formwork striking times
- CIRIA R108 (1985) Concrete pressure on formwork
- CIRIA SP57 (1988) Handling of materials on site

Background

- CIRIA SP118 (1995) Steel reinforcement

Classification

CAWS Group E, **CEWS** Class F, **CI/SfB** Code (2-)E

Group C – Primary structure
3 – Pre-cast concrete

Read the Introduction before using the following guidelines

Scope

- Pre-cast work on site – cast adjacent to final location.
- Pre-cast work offsite – resultant effect of pre-cast construction decisions on the site.
- Transportation – from off-site pre-cast yard to construction site.
- Lifting – during fabrication, transportation, installation.
- Positioning – setting out, tolerances, locating, securing.
- Bearings/supports – adequacy, stability, temporary works.
- Fixings and holes/openings for handrails, ladders, barriers, services, chases.
- Finishes – to avoid site preparatory activities.
- Temporary loading, supporting, propping, bracing.

Exclusions

- Railway works (see E3).
- Marine works.
- Pre-stressed concrete (see C4).
- Cladding panels (see D1).

Major hazards

Refer to the Introduction for details of accident types and health risks

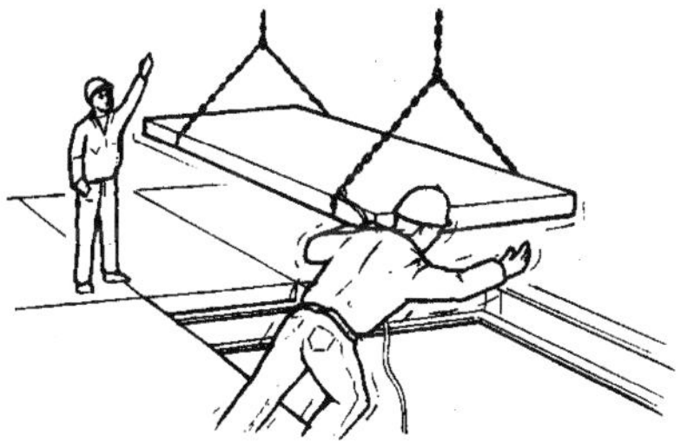

Falling from height
Operatives falling while accessing, locating or fixing pre-cast units.

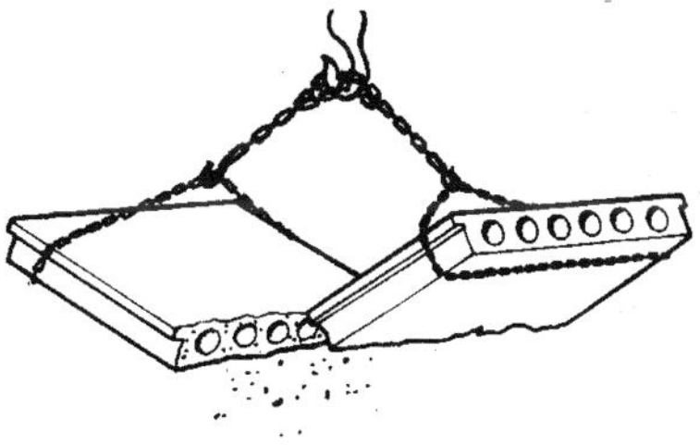

Crushing
Lifting procedure for unit not defined in design:
- adequate lifting points
- maturity of concrete at lifting.

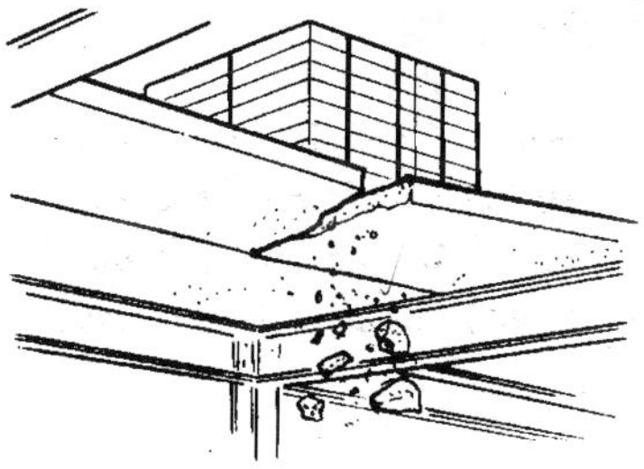

Collapse:
Pre-cast unit unable to carry loading or insufficiently restrained in temporary condition.

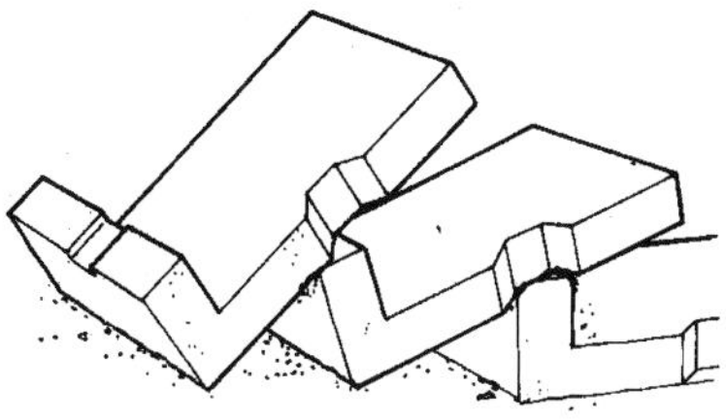

Collapse
Units placed in wrong position (inverted or on their side or located with incorrect bearing).

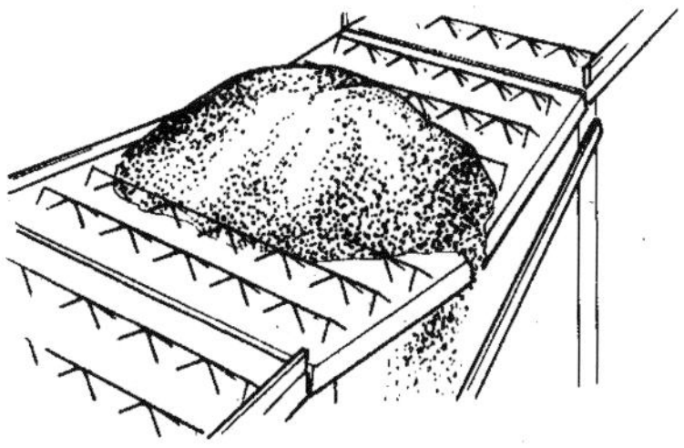

Collapse:
Pre-cast composite units may fail if not propped to carry the in-situ topping and temporary load.

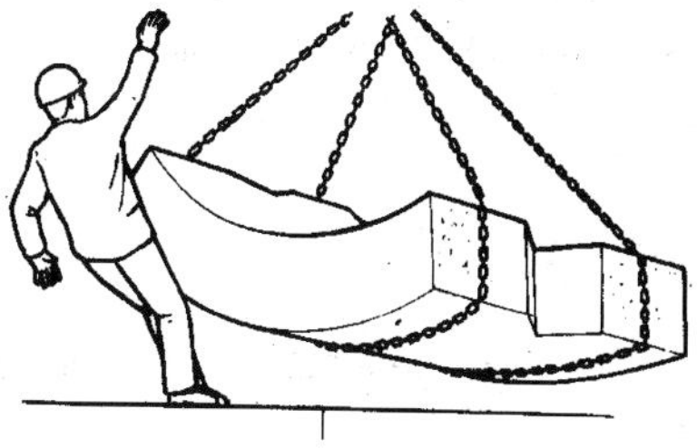

Handling
Pre-cast unit is a shape/weight that is difficult to transport, lift or position.

C3 Group C – Primary structure 3 – Pre-cast concrete

Specific hazard identification

Possible key considerations:

- **Tolerance control** – encourages rapid and safe assembly.
- **Work at height –** Where are critical considerations?
- **Temporary conditions –** Are the assumed temporary conditions set down? Show that it is designed for fabrication, lifting, transport placing, temporary loading, etc.
- **Installation –** Do the units require installation procedures that are not likely to be routine on site?
- **Procedures** – Ensure that the procedures described for unit installation are sufficiently comprehensive.
- **Insitu stitch –** Where are details of in-situ infills or topping?
- **Any others?**

Prompts

- Access/transport
- Mock-up
- Handling points
- Unit weights
- Lifting procedures
- Transporting
- Positioning
- Temporary support
- Composite action
- Orientation
- Demolition
- Training

Hazards consideration in design

Stage	Considerations and issues	Possible design options to avoid or mitigate hazards identified for pre-cast concrete
Concept design	• Why pre-cast? • Scope of work • Familiarly used? • Define accessibility • Interfaces set out • Sequence defined • Overall mass • Technology known	• Do the pre-cast concrete elements form an insignificant part of the work? Are they to be cast on site? If so, try to ensure details are simple. • Clearly define the designers' intended construction sequence • Are the units large? Can they be easily transported and handled and positioned on site? • Can pre-cast units be easily and safely positioned in their likely location?
Scheme design	• Pre-stressed? • Unit characteristics • Span/section known • Shape identified • Number of units • Loading condition • Protection defined • Tolerances set out • Procedures defined	• Is the pre-cast supplier likely to have made similar units previously? • Ensure that the limits of size, weight and number are known by all parties as soon as possible. • Specify how the units are to be protected, handled, lifted and positioned. Include lifting points. • Ideally the units should be lifted and finally located direct from their transporter.
Detail design	• Handling designed • Handling formwork and reinforcement • Surface treatment • Cutting/sawing • Inserts/fixings • Positioning/locate • Temporary support	• Ensure that all holes, chases, conduits, fixings, etc. are cast into the precast unit. • Ensure that detailing of reinforcement cages provides for easy lifting. Identify lifting points. • Minimise and simplify locating and fixing procedures and details especially at height or depth.

Examples of risk mitigation (methods of solution)

ACTION \ ISSUE	Element design	Connections/tolerances
Avoidance Design to avoid identified hazards but beware of introducing others	Standardise unit size and shape.	Simplify unit shape for tolerance control.
Reduction Design to reduce identified hazards but beware of increasing others	Consider limit for unit size/shape/weight.	Simplify lifting fixing/connection details.
Control Design to provide acceptable safeguards for all remaining identified hazards	Detail unit handling procedures/training.	Specify location/ connection procedures.

Examples of risk mitigation (issues addressed at different stages)

Apartment blocks on a site are required to be built within a short site construction period. To achieve the programme and budget, it has been decided to maximise the off-site fabrication process. One option is the use of pre-cast concrete for structural elements.

Concept design	Scheme design	Detail design
Pre-cast construction, judged to allow greater standardisation – avoids hazards of complicated in-situ details and work on site.	Standardise units where possible. Consider dimensions, weight and shape of units. Design for loads during lifting.	To reduce handling and fixing hazards, the pre-cast units will include cast-in connections for handling, positioning, hand rails and running lines

An office block is to be built over vaulted arches which are to be retained. The columns for the offices pass through the vaulting and are founded below the vaults. Due to lack of headroom in the vaults and high loading, the foundations are proposed to be hand-dug caissons with an option to use pre-cast segmental rings as liners.

Concept design	Scheme design	Detail design
This application of caissons is not a familiar technique. Designers are examining the pre-cast liner option in order to avoid hazards of collapse.	A design study determines that a 3 segment pre-cast concrete liner reduces hazards over the in-situ option.	Mock-up trials indicated pre-cast liner segments should include lifting sockets, grout/water seals, grouting and vent tubes and escape ladder fixings.

A GENERAL PLANNING
B EXCAVATIONS AND FOUNDATIONS
C 3 PRIMARY STRUCTURE
D BUILDING ELEMENTS AND BUILDING SERVICES
E CIVIL ENGINEERING

Related issues

References within this document

References	Related issues
A1	Surrounding environment will need to be considered with regard to transporting and positioning pre-cast units.
A2	Site clearance may be required in order to prepare for access and positioning of pre-cast units.
A3	Site investigation could be required in order to finalise which elements are pre-cast.
A4	Access will be required for the transportation, positioning and placing (or removal) of the unit.
A5	Site layout will interact with access, positioning and placing of pre-cast units.
B1	General excavations could entail the use of pre-cast units. Ensure they will resist the conditions.
B2	Deep basements/shafts may use pre-cast units as lining in segmental form for retaining structures.
B3	Trenches for foundations and services, pre-cast units may be used to carry services.
B4	Retaining walls – may be pre-cast concrete.
B6	Piling – may use pre-cast concrete for driven piles.
C1	General concrete – ensure the pre-cast units perform adequately.
C2	In-situ concrete may be considered as an alternative for pre-cast concrete.
C4	Pre-stressed, post-tensioned concrete may be pre-cast concrete.
D1	External cladding likely to be pre-cast if it is to be made of concrete.
E1	Civil engineering, small works will frequently use pre-cast concrete units.
E4	Bridge construction may be entirely or partly pre-cast concrete.
E5	Bridge maintenance could entail the maintenance of pre-cast concrete units and the sealing, joints etc.
E6	Working or near water could use pre-cast concrete.

References for further guidance

Primary general references and background information given in Section 4

Primary

- CIRIA TN137 (1991) Selection and use of fixings in concrete and masonry (microfiche)
- CIRIA SP57 (1988) Handling of materials on site
- Precast Concrete Frame Association: Code of practice for safe erection of pre-cast concrete frameworks (1999)

Secondary

- BS 5975:1996 Code of practice for falsework
- CIRIA TN104 (1981) Pre-cast concrete tunnel linings
- HSE GS28 Parts 1 to 4 Safe erection of structures

Background

- BS 7121: Code of Practice for safe use of cranes
  - Part 1:1989 General
  - Part 2:1991 Inspection, testing and examination
- CIRIA C703 (2003) Crane stability on site. Second edition

Classification

CAWS Group E, **CEWS** Class H, **CI/SfB** Code (2-)f

Group C – Primary structure
4 – Pre-stressed, post-tensioned concrete

Read the Introduction before using the following guidelines

Scope

- Post-tensioned reinforced concrete.
- Reinforcement – span, anchor zones, duct supports.
- Stressing ducts – gauge, profile, fixing, jointing.
- Stressing cables – wires, strands.
- Anchorage zones – anchor blocks, wedges, jacks.
- Grouting – grouting tubes, venting, duct profiles.
- Stressing procedure – when, sequence, linear, monitor, precautions.
- Materials – concrete maturity, strength, creep, shrinkage.
- Demolition.

Special notes:

1. Stresses used in pre-stressed concrete are usually very high. Care must be taken in design and use of all materials
2. Wire = single
   Strand = multi-wire
   Tendon = multi strand

Exclusions

- Pre-stessed pre-tensioned concrete.
- Externally post-tensioned structures eg silos, bridges.
- Pre-stressed non-reinforcement concrete structures eg. ground anchors.

Major hazards

Refer to the Introduction for details of accident types and health risks

Falls from height
Operatives fall from access/ elements during fixing/ stressing.

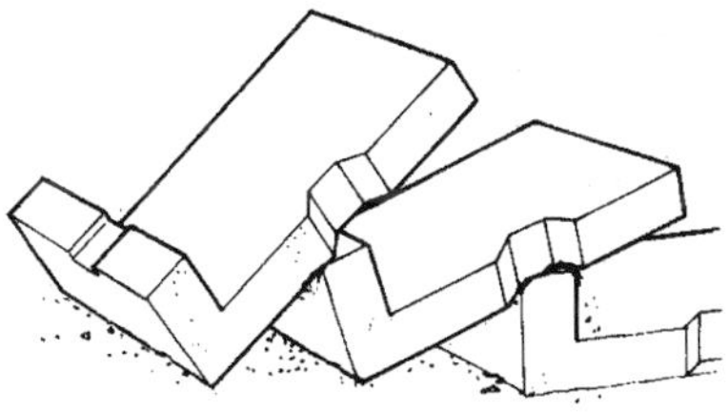

Collapse/crushing
Units unstable in temporary condition (high d/b ratio).

Impact/collapse
Strand breakages during stressing:
- strand damage or duct obstruction/defect
- duct alignment.

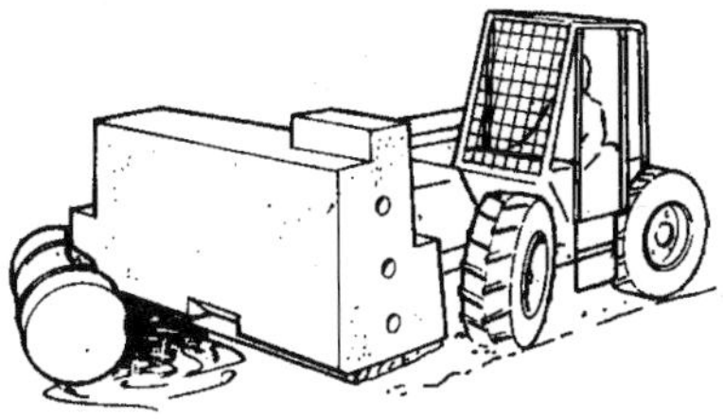

Inadequate access
Operatives unable safely to access work area, poor materials handling facilities.

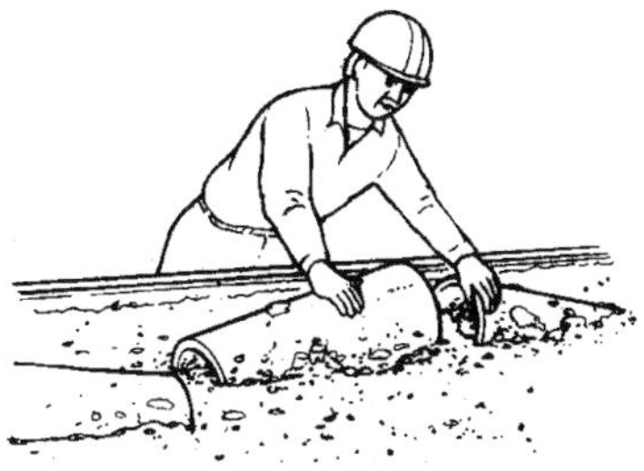

Collapse
Duct breaks loose and floats during concreting
(duct support failure).

Impact/collapse
Collapse of element during demolition. Inadequate support and failure of compression zone during demolition.

Group C – Primary structure 4 – Pre-stressed, post-tensioned concrete

Specific hazard identification

Possible key considerations:

- **Assumptions** – Where are the design assumptions stated clearly?
- **Materials** – How will materials be maintained in good condition on site?
- **Access** – Show that the design will allow sufficient working space.
- **Design** – Show that the design avoids strand, anchor, reinforcement congestion.
- **Support** – Provide adequate support for stressing ducts.
- **Grouting** – Check grout vents are to be at highest point of duct profile.
- **Ducts** – How are stressing ducts kept clear of obstructions and water?
- **Strands** – Ensure strands are not snagged in ducts.
- **Any others?**

Prompts

- Protect materials
- Access for fixing
- Access for stressing
- Stressing protection
- Grouting procedure
- Handling points
- Transportation
- Demolition

Hazards consideration in design

Stage	Considerations and issues	Possible design options to avoid or mitigate hazards identified for pre-stressed concrete
Concept design	• Why pre-stress? • Scope of work • Plan accessibility • Define interfaces • Sequence assumed • Overall mass • Technology known?	• Pre-stressing can be hazardous if inadequate design guidance to the contractor. • Loads and stresses are high – more easily controlled in a factory? Are the elements to be made on site? • Will the necessary construction expertise be locally available to the site?
Scheme design	• Unit characteristics • Span section review • Number of units • Loading conditions • Locate tendons • Anchorage access • Duct support • Stressing protection • Tolerances defined	• Units should be standardised wherever possible. • Prepare the design so that it allows the work to be carried out on site in good working conditions. • Ensure that the design sequence allows the contractor to provide good access. • Design and detail to reduce the risk that the materials will be damaged prior to or during installation (eg to reduce ad hoc repair work on site).
Detail design	• Procedures defined • Care of materials • Tendon layout • Duct profile, radii, joints, vents • Anchorage layout • Strands replaced • Stressing sequence • Grouting sequence • Striking supports • Deflection/creep • Loading application	• Specify conditions for materials storage, dry, well laid-out, clean, secure, etc. • Be aware of strand installation method when designing ducts. • Design duct supports. • Ensure adequate spacing for anchorages. • Avoid dead-end anchorages. • Specify installation procedures and controlled inspections of work. • Specify protection of materials, installation and completed work. • Detail control procedures prior to, during and after tendon stressing to reduce risk of hazards occurring.

Examples of risk mitigation (methods of solution)

ACTION / ISSUE	Reduce and simplify site activities	Cable failure during post-tensioning
Avoidance Design to avoid identified hazards but beware of introducing others	Pre-construct off-site or at ground level.	Avoid stressing duct congestion.
Reduction Design to reduce identified hazards but beware of increasing others	Pre-stressed proposals simplified and good access provided.	Smooth stressing duct profile.
Control Design to provide acceptable safeguards for all remaining identified hazards	Outline competence level required and highlight the need to co-ordinate with concurrent trades.	Outline stressing procedures.

Examples of risk mitigation (issues addressed at different stages)

A building constructed in the 1930s has undergone change of use to offices. This requires a review of the existing rivetted steel frame.

Concept design	Scheme design	Detail design
To carry additional loading, it requires long span plated steel beams to be upgraded. The beams will be cased in reinforced concrete and post-tensioned in order to carry the new floor loading.	The composite post-tensioned solution avoids hazards of gas cutting and welding, while minimising noise and dust	Provide relevant information to Planning Supervisor for sequence of working and design assumptions.

The ramp beams are to be built for an elevated road deck. Due to alignment, the elevated ramp structure is complex and the design needs to reduce the hazards of congested construction at height.

The geometry of the beams are complex due to curved alignment on plan and elevation. The beam depth will be restricted by headroom constraints below them.

Concept design	Scheme design	Detail design
The ramp beams for an elevated road deck are to be cast in-situ post-tensioned concrete, instead of pre-cast, in order to avoid difficult details of casting PC units, but will now involve increased work at height.	The in-situ design has avoided the hazards associated with complex layout of stressing tendons.	To reduce hazards, details provide adequate spacing of anchor blocks in relation to stressing jack requirements.

Group C – Primary structure
4 – Pre-stressed, post-tensioned concrete

A GENERAL PLANNING
B EXCAVATIONS AND FOUNDATIONS
C 4 PRIMARY STRUCTURE
D BUILDING ELEMENTS AND BUILDING SERVICES
E CIVIL ENGINEERING

Related issues

References within this document

References	Related issues
A1	Surrounding environment may have to be considered for stressing procedures and protection.
A2	Site clearance and simple demolition will need to identify those elements that are post-tensioned.
A3	Site investigation may be required to determine the extent and details of pre-stressed concrete within a structure.
A4	Access on the site will need to consider location and access to pre-stressed units.
A5	Site layout will need to identify the positioning of pre-stressed units, to design access and protection.
B4	Retaining walls may be designed with the use of pre-stressed concrete, (pre-tensioned or post-tensioned).
B5	Ground stabilisation may be carried out with the use of ground anchors (in effect, a form of post-tensioned concrete).
C1	General concrete will identify the constituents of the concrete in the pre-stressed member.
C2	In-situ concrete – Post-tensioned unit is more likely to be cast in-situ (in position).
C3	Pre-cast concrete – The post-tensioned units may be pre-cast prior to final location.
D7-D10	Services – Extreme caution should be exercised if services require the formation of builders work openings in pre-stressed units. They should all be pre-formed prior to concreting. Never cut on site.
E1-E8	Civil engineering works – Pre-stressed concrete frequently occurs in civil engineering works.
E4	Bridge construction is often pre-stressed concrete.
E5	Bridge maintenance should always check pre-stressed members during condition surveys.
E6	Working over/near water may use pre-stressed concrete.

References for further guidance

Primary general references and background information given in Section 4

Primary

- HSE GS49 Pre-stressed concrete
- CIRIA SP57 (1988) Handling of materials on site

Secondary

- CIRIA R28 (1971) The connection of pre-cast concrete structural members (microfiche)
- CIRIA TN129 (1987) Prestressed concrete beams – Controlled demolition and pre-stress loss assessment
- HSG 150 Health and safety in construction
- HSE INDG258 Safe work in confined spaces

Background

- BS 5975:1992 Code of practice for falsework

Classification

CAWS Group E, **CEWS** Class H, **CI/SfB** Code (2-)E

Group C – Primary structure
5 – General steelwork

C5

Read the Introduction before using the following guidelines

Scope

- Building frames such as offices, schools, hotels, etc.
- Portal frames for agricultural or industrial premises.
- Towers and chimneys, assembled on site.
- Footbridges and other load carrying spans.
- Gantries for street signs/lighting and other similar.
- Atria or other facades based on a structural steel frame.
- Hot-rolled and cold-formed sections.

Exclusions

- Temporary works.
- Stability during erection (see C6).
- Corrosion/fire protection (see D5).
- Stressed skin designs.
- Cable tied and fabric structures.

Major hazards

Refer to the Introduction for details of accident types and health risks

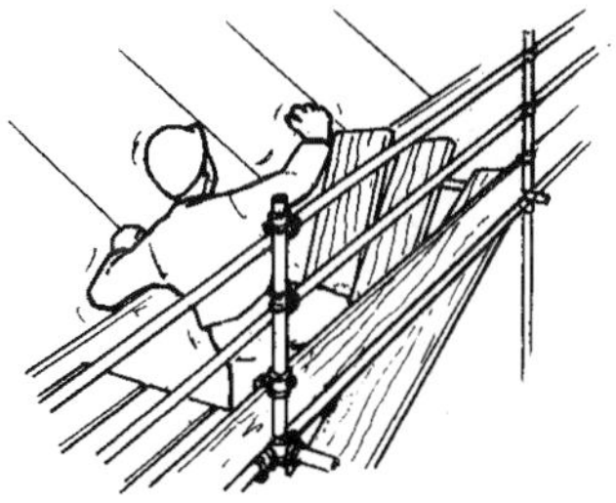

Falls from Height
Placing, stability and provision of working platforms, access.

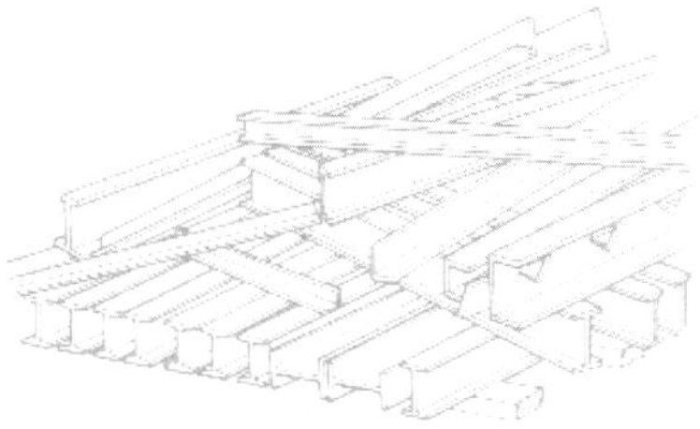

Collapse
Storage areas, stacking and stability, falling objects, access, temporary bracing.

Hazardous operations
Cutting/Welding.
Painting.
Descaling.

Mobile Plant
Delivery, traffic control, offloading, slinging.
Site routes.

Moving objects
Remedial works, plumbing, levelling.
Trapping.
Manual handling.

Noise, vibration
Bolting, drilling.
Reaming.

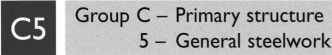

C5 Group C – Primary structure
5 – General steelwork

Specific hazard identification

Possible key considerations:

- **Delivery** – How will pieces be manoeuvred to and onto site?
- **Logistics** – How will delivery match sequence of use?
- **Lifting** – What are the requirements to allow the (tower) crane driver the facility to lift and position the pieces safely?
- **Storage** – Where and how will items be stored as they are unloaded?
- **Stability** – When unloaded, how will stability of the pieces be controlled?
- **Access** – How will hand and power tools be used at the work site?
- **Working at height** – How will working at height be accomplished?
- **Light and air** – What sorts of local environments are involved?
- **Other activities** – How could the works affect other people? How might the workforce be affected by what is going on around them?
- **Any others?**

Prompts

- Lifting
- Access
- Transport
- Positioning
- Alignment and fit-up
- Reaming
- Slotted holes
- Connection type
- Corrosion protection
- Welding
- Cutting
- Bolting
- Painting

Hazards consideration in design

Stage	Considerations and issues	Possible design options to avoid or mitigate hazards identified for general steelwork
Concept design	• Why use steelwork? • Scope of work • Element sizes • Piece weights • Manoeuvrability • Equipment	• Exploit the facility to quickly erect and achieve stability; a safe working platform offers specific safety advantages. • Determine how the steel frame can provide working platforms for subsequent work and other trades. • Assess methods for delivery, storage, pre-assembly and erection of steelwork to safeguard construction workers. • Metal decking for composite slabs offers significant safety advantages during construction (eg stabilises the structure, provides safe working platform).
Scheme design	• Site Environment • Programme • Craneage • Working area • Sub assemblies • Movement joints • Temporary loads • Interface connections • Sequencing of work	• Show how structure might be built to anticipate and eliminate (if possible) hazards. • Consider how temporary/erection loads can be accommodated without the need for additional members or framing. • Explore how pre-assembly off and on site can enhance health and safety. • Consider installing staircases earlier to provide safer access. • Ensure adequate bracing is provided as the frame is erected
Detail design	• Connection types • Bolt types • Temporary conditions • Pre-assembly • Lining and levelling • Tolerances • Painting • Testing	• Consider advantages of repetition (eg one size, fully-threaded bolts) to reduce complexity of hazardous work. • Ensure that connections are dimensionally tolerant to avoid need for remedial works at height. • De-burr holes and grind arrises to make components safer to handle. • Eliminate/reduce need for work on site at height by ensuring an adequate base is prepared to allow safe working of mechanical lifting plant. • Avoid complex connections.

Examples of risk mitigation (methods of solution)

ACTION \ ISSUE	Moving and storing materials	Working at height
Avoidance Design to avoid identified hazards but beware of introducing others	Place steelwork straight from the lorry	Consider pre-assembly on the ground
Reduction Design to reduce identified hazards but beware of increasing others	Establish site to provide adequate laydown areas away from site traffic	Design-in lifting points and/or methods to release crane hooks remotely.
Control Design to provide acceptable safeguards for all remaining identified hazards	Use guys and timbers to support steelwork until placed in position	Incorporate safety harness attachment points in detailing

Examples of risk mitigation (issues addressed at different stages)

For a large portal frame warehouse, the designer wants to provide access within the building envelope for other works while the steel frame is being erected and before the building is clad and finally braced.

Concept design	Scheme design	Detail design
A sequence of construction is designed so that other trades can rapidly follow on behind steelwork erection; bracing provided so that structural stability always maintained.	A method of pre-assembly and erection is established to ensure that the contractor can be advised of a way in which it can be built safely.	Baseplates and foundation are detailed to provide adequate resistance to overturning, by using four bolts to accommodate temporary conditions.

A building refurbishment project involves the retention of a listed facade and the reconstruction of part of the building to link up to a 1930s steel-framed structure.

Concept design	Scheme design	Detail design
Member serial sizes and lengths are kept down to a size which can be "threaded" into the structure without requiring a tower crane.	Bolted connections are adopted throughout because of the problem of welding in confined spaces with a lot of adjacent activity.	Connections are designed to be dimensionally tolerant, so that risk of misfit is minimised.

Group C – Primary structure
5 – General steelwork

A GENERAL PLANNING
B EXCAVATIONS AND FOUNDATIONS
C 5 PRIMARY STRUCTURE
D BUILDING ELEMENTS AND BUILDING SERVICES
E CIVIL ENGINEERING

Related issues

References within this document

References	Related issues
A1	Surrounding environment will need to be considered when examining access to deliver steelwork.
A2	Site clearance and simple demolition may be needed so that steelwork storage and placing activities can proceed.
A4	Access on the site will need to be determined to ensure the safe construction.
A5	Site layout may interact with the decision to use steelwork, particularly if pre-assembly is used more generally.
C1	General concrete is usually used for foundations for steelwork.
C6	Temporary stability is a dominant issue when considering safe design in steelwork.
D3	Atria are invariably designed in steelwork with glazing infill.
D5	Surface coatings and finishes will always be required on exposed steelwork.
D10	Lifts and travelators could have steel framing or have steelwork linking it to a concrete core.
E4	Bridge construction may be steel and always has health and safety considerations.
E5	Bridge maintenance will probably require periodic attention to steelwork.
E6	Working over/near water may use any steelwork.
E7	Cofferdams in water may use steelwork.

References for further guidance

Primary general references and background information given in Section 4

Primary

- HSE GS28 Parts 1 to 4 Safe erection of structures
  - Part 1 Initial planning and design
  - Part 2 Site management and procedures
  - Part 3 Working places and access
  - Part 4 Legislation and training
- CIRIA C703 (2003) Crane stability on site. Second edition
- CIRIA SP57 (1988) Handling of materials on site

Secondary

- British Constructional Steelwork Association (1999) Safer erection of steel frame buildings
- British Constructional Steelwork Association (1980) Structural steelwork fabrication
- Steel Construction Institute (2003) Steel Designers' Manual (Blackwell)
- CIRIA R87 (1981) Lack of fit in steel structures

Background

- BS 7121 Code of Practice for safe use of cranes
  - Part 1:1989 General
  - Part 2:2003 Inspection, testing and examination

Classification

CAWS Group G, **CEWS** Class M, **CI/SfB** Code (28)h2

Group C – Primary structure
6 – Stability and erection of structural steelwork

C6

Read the Introduction before using the following guidelines

Scope

- Site pre-assembly.
- Pre or part erection.
- Temporary works.
- Interim connections.
- Erection stability.
- Plumbing and levelling.
- Temporary bracing.
- Structural cores.
- Interfaces with other trades.
- Connections.

Exclusions

- Temporary storage or laydown.
- Launched structures.
- Stressed skin designs.
- Cable tied and fabric structures.
- Demolition.

Major hazards

Refer to the Introduction for details of accident types and health risks

Falls from height
Placing, stability and provision of working platforms, access.

Collapse
Excessive loading or load reversal.
Inadequate ties, struts or bracing.
Incorrect sequence.
Incomplete assembly.

Noise vibration
Pneumatic and manual wrenches and hammers.

Harmful Substances
Sparks and heat from burning and welding.
Electricity
Paints, greases etc.

Moving objects/trappings
Placing of steelwork. Closure of joints.
Sharp objects.
Baseplate packing.

Refer also to Sector C5 hazards

Specific hazard identification

Possible key considerations:

- **Handling** – How are units to be brought together/ assembled?
- **Assembly –** What different methods of assembly are safest for this site?
- **Buildability –** Determine at an early stage how the structure might be built safely. Repeatedly reconsider this as design proceeds to ensure that safety is maintained or enhanced.
- **Structural Capacity –** Which methods and sequences of assembly could inadvertently lead to overloading?
- **Stability Core** – If any parts of the structure rely upon other particular parts for their stability, is this obvious?
- **Site Condition** – What site conditions could affect these works? How can these be designed out?
- **Pieces (and bundles) of steel** could have their weight marked and locations and orientations shown, making placement easier.
- **Manoeuvring –** Once on the crane hook, do loads need additional control? How will this be provided?
- **Any others?**

Prompts

- Local access
- Visibility
- Ground stability
- Surrounding environment
- Confined spaces
- Temporary guys/struts
- Alignment
- Special connections
- Corrosion protection
- Welding
- Cutting
- Loose fit pieces
- Cleaning and painting
- Checking and testing
- Base grouting

Hazards consideration in design

Stage	Considerations and issues	Possible design options to avoid or mitigate hazards identified for stability and erection of structural steelwork
Concept design	• Scope of work • Form of construction • Building shape/size • Pre-assembly • Sequences • Major Plant	• Because the highest incidence of accident and injuries arise from "falls from height" this must be a special consideration when considering ways to allow safe work on partially-erected steel frames. • Consider sequence of site/structure development to determine the most appropriate option.
Scheme design	• Site access • Craneage • Ground conditions • Work sequences • Sub-assemblies • Element sizes/weights • Temporary loads • Work platforms	• Can the steel frame erection be conceived as inherently stable sub-assemblies? • Consider if welding, HSFG, or bolts in bearing offer any particular health and safety benefits – eg by providing temporary moment capacity. • Assess structure to see if areas can be rapidly developed to act as stable work platforms. • Design so that stairways can be installed ahead of frame erection, for safer access.
Detail design	• Inter-connections • Connection details • Lot sizes and sequences • Column bases • Temporary storage • Bolt types and sizes • Tightening and torquing • Tolerances	• For ease of erection, seek to use fewer – or even one – size of bolt throughout. • Can elements in tension also be designed to accept temporary compression loads (to provide enhanced stability)? • Ensure that structural mechanism assumed for method of construction is communicated to all who might need to know. • Include temporary working platforms and/or harness attachment points.

Group C – Primary structure
6 – Stability and erection of structural steelwork

Examples of risk mitigation (methods of solution)

ACTION \ ISSUE	Assembly stability	Making connections
Avoidance Design to avoid identified hazards but beware of introducing others	Design for assembly at ground level and to have fewer but larger lifts. Ensure provision of base for cranage.	Adopt positive seating systems. Explore range of connection types.
Reduction Design to reduce identified hazards but beware of increasing others	Outline sequence of construction which allows in-built stability.	Specify one size, fully-threaded free running bolts to simplify work at height.
Control Design to provide acceptable safeguards for all remaining identified hazards	Outline temporary support.	Allow for use of hydraulic platform for jointing at height.

Examples of risk mitigation (issues addressed at different stages)

Because the top three storeys of a building cantilever out over other structures, the engineer has arranged that they structurally "hang" from a central core in the final condition. During erection, however, the steel frame can be sufficiently light as to be temporarily supported from the outrigger areas of the floors below.

Concept design	Scheme design	Detail design
Establish work platform/crash deck areas and deal with tower crane oversailing to ensure practicality	Identify construction sequence to allow structure to develop on a floor-by-floor basis.	Design "hangers" also for temporary role as columns, to ensure stability at all stages of erection.

A steel footbridge is to be erected over a busy highway which serves a housing estate, a school and a shopping precinct.

Concept design	Scheme design	Detail design
A single span is chosen, to minimise working adjacent to traffic and to reduce dangers to road vehicles.	A lightweight design is adopted so that it can be lifted into place during a single weekend night-time road closure. Minimum temporary works required for temporary stability.	Connections and lifting points in the steelwork are designed to allow rapid pre-assembly adjacent to the site.

Group C – Primary structure
6 – Stability and erection of structural steelwork

Related issues

References within this document

References	Related issues
A1	Surrounding environment will need to be considered when examining access to site.
A2	Site clearance and simple demolition may be relevant for steel erection.
A3	Site investigation may have to be carried out before design of the column bases.
A4	Access on the site will need to be determined to ensure safe construction .
A5	Site layout may affect the decisions to use steelwork.
C1	General concrete.
C5	General steelwork.
D3	Atria – Invariably have temporary and final stability issues to be resolved.
D5	Surface coatings and finishes – will always be part of erecting steelwork.
D10	Vertical transportation – often a part of the stability core of a structure.
E4	Bridge construction may well be constructed using steel.
E5	Bridge maintenance will probably require periodic attention to steelwork details.
E6	Working over/near water will be likely to use steelwork.
E7	Cofferdams in water may use steel.

References for further guidance

Primary general references and background information given in Section 4

Primary

- BS 5531:1988 Code of practice for safety in erecting structural frames
- HSE GS28 Parts 1 to 4 Safe erection of structures
- CIRIA SP131 (1996) Crane stability on site

Secondary

- British Constructional Steelwork Association (1999) Safer erection of steel frame buildings
- British Constructional Steelwork Association (1980) Structural steelwork fabrication
- Steel Construction Institute (2003) Steel Designers' Manual (Blackwell)
- CIRIA SP57 (1988) Handling of materials on site

Background

- BS 7121: Code of Practice for safe use of cranes
  - Part 1:1989 General
  - Part 2:2003 Inspection, testing and examination

Classification

CAWS Group G, **CEWS** Class M, **CI/SfB**, Code (28)h2

Group C – Primary structure
7 – Masonry

GENERAL PLANNING A
EXCAVATIONS AND FOUNDATIONS B
PRIMARY STRUCTURE C 7
BUILDING ELEMENTS AND BUILDING SERVICES D
CIVIL ENGINEERING E

Read the Introduction before using the following guidelines

Scope

- Brickwork.
- Blockwork.
- Stonework.
- Pre-assembled composite masonry/brickwork units.
- Jointing materials.
- Positioning of pre-assembled elements including setting out, locating and securing.
- Bearings/supports of pre-assembled elements.
- Temporary loading and supporting.
- Reinforcement in masonry work.

Exclusions

- Masonry cladding (see D1).
- Gabions/crib block walls.
- Post-tensioned masonry.

Major hazards

Refer to the Introduction for details of accident types and health risks

Handling
Weight, shape, transportation of units. Working space, hoists, manual handling at height.

Hazardous substances
Cement handling, lime and mortar mixing, chemical grouts, additives, dust.

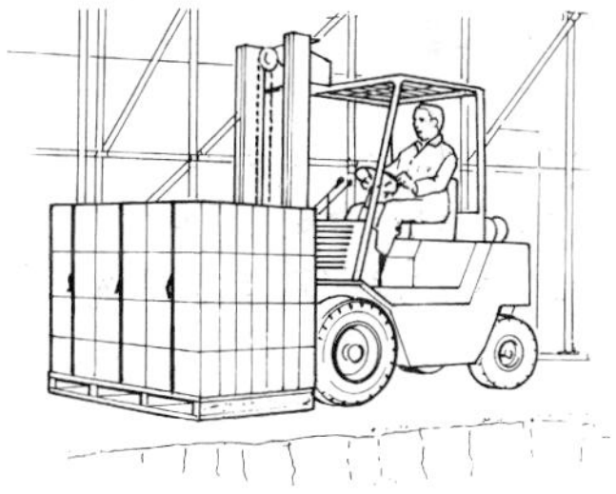

Access
Delivery/offloading. Plant on site. Obstruction by other building elements. Working space restrictions.

Crushing
Placement of units, insufficient space for storage/working, falling objects, vehicle movements.

Collapse
Unpropped construction, elements being overloaded, movement joints, storage/stacking irregular units, horizontal chases, DPC debonding.

Abrasion/cuts
Masonry texture, wall ties, metal lathing, sheet metal
Insulation (glass fibre).

Specific hazard identification

Possible key considerations:

- **Handling** – How will the units be handled? Manually or by crane?
- **Access** – What provisions are required to ensure adequate space for plant and scaffolding?
- **Temporary conditions** – What design considerations are involved?
- **Installation** – How can this be arranged to ensure healthy and safe procedures?
- **Special considerations** – What are the required procedures (eg cutting)?
- **Jointing materials** – What are these to be? How will they be applied?
- **Masonry ties** – What can be specified to make installation safer?
- **Any others?**

Prompts

- Source of materials
- Handling units
- Handling points
- Lifting procedures
- Transporting units
- Positioning
- Temporary support
- Lime and other chemicals
- Cleaning
- Demolition

Hazards consideration in design

Stage	Considerations and issues	Possible design options to avoid or mitigate hazards identified for masonry
Concept design	• Why use masonry? • Location • Scope of work • Familiar work • Access • Interfaces • Sequences • Materials	• The alternatives to using masonry will generally involve in-situ or pre-cast concrete. If not, design masonry that has familiar details. • Can stone work be factory prepared? In particular, can site cutting of sandstone be avoided? • Plan the timescale and sequence of masonry working to avoid working during bad weather, during winter months or on a congested site. • Study alternative routes for access. Allow adequate working space for operatives and materials.
Scheme design	• Unit types • Joints • Handling • Reinforcement • Protection • Tolerances • Procedures • Working space	• Specify materials that are least likely to create dust (eg use preformed materials). • Avoid blocks above 20kg weight. • Consider components that are preformed and joined on site. • Consider methods of quick and secure erection. • Joint design tolerances should be relevant to nature of work. • Design provision for delivery and handling of materials on congested sites, close to work station if possible.
Detail design	• Handling • Inserts/fixings • Connections/joints • Finishes • Bearings • Positioning • Temporary supports	• Wherever possible use materials that are non-corrosive and non-abrasive. • Specify appropriate tolerances to avoid cutting, etc. on site. • If mechanical handling is necessary, incorporate lifting and handling attachment points into the design. • Design-in provision for service runs, to avoid cutting on site.

Examples of risk mitigation (methods of solution)

ACTION \ ISSUE	Handling large blocks	Use of brickwork ties
Avoidance Design to avoid identified hazards but beware of introducing others	Use blocks less than 20kg, to allow manual handling.	Specify ties without sharp edges.
Reduction Design to reduce identified hazards but beware of increasing others	Detail lifting points for block handling.	Construct second skin as soon as possible.
Control Design to provide acceptable safeguards for all remaining identified hazards	Specify mechanical handling for laying large blocks.	Specify temporary covers to sharp ties (if used).

Examples of risk mitigation (issues addressed at different stages)

A refurbishment project requires selected stonework to be removed and replaced at height from a building due to weathering and moisture ingress. The new stonework is to match the existing.

Concept design	Scheme design	Detail design
Consider alternative safe methods for how the suspect stonework could be removed and replaced.	Develop a safe method of removal and replacement using scaffolding as access and also means of handling stonework.	To avoid hazards of cutting on site, special replacement stonework will be surveyed and prepared off-site.

Internal fair-faced walls to an exhibition centre are 8 metres high and are required to have the special effect of textured large concrete blocks with decorative joints.

Concept design	Scheme design	Detail design
Design for 2 leafs of blockwork, using 20kg blocks to reduce weight for manual handling.	Design 20kg blocks to incorporate provisions for the decorative joints.	Blockwork lifts not to exceed 1.5m high each day. Wall-ties selected to avoid hazards of handling and being cut by sharp edges. Highlight that care is required when using retarders.

C7

Group C – Primary structure
7 – Masonry

Related issues

References within this document

References	Related issues
A2	Site clearance and simple demolition will involve the safe storage and removal of brickwork and general masonry.
A4	Access onto site will need to accommodate the frequent deliveries of materials for masonry work.
A5	Site layout will need to consider the safe storage of materials delivered to site for masonry work.
B1-B3	Excavations are frequently for masonry work within deep excavations. Alternative methods should be considered (eg use of pre-cast rings).
B7	Underpinning to foundations could involve masonry work.
C1-C4	Concrete works usually interface with masonry works, with different trades working closely.
D1	External cladding often incorporates masonry panels.
E1	Civil Engineering small works may involve the use of engineering brickwork/masonry work, in difficult and remote locations.

References for further guidance

Primary general references and background information given in Section 4

Primary

- BRE Good Building Guide 1 Repairing and replacing lintels
- HSE CIS36 Silica
- HSE CIS37 Handling building blocks
- CIRIA R117 (1988) Replacement ties in cavity walls: a guide to tie spacing and selection
- CIRIA SP57 (1988) Handling of materials on site
- CIRIA C589 (2003) Retention of masonry facades – best practice site handbook

Secondary

- BRE Good Building Guide 14 Building simple plan brick or blockwork freestanding walls
- BRE Defect action sheet 115 External masonry cavity wall, wall ties, selection and specification

Background

- Orton, A (1992) Structural design of masonry (Longman)
- BS 5628: Code of practice for use of masonry
  - Part 1:1992, Structural use of unreinforced masonry
  - Part 2:1995, Structural use of reinforced and prestressed masonry
  - Part 3:1985, Materials and components, design and workmanship

Classification

CAWS Group F, **CEWS** Class U, **CI/SfB** Code (2-)F

Group C – Primary structure
8 – Timber

Read the Introduction before using the following guidelines

Scope

- Softwoods.
- Hardwoods.
- Wood-based products (eg plywood, chipboard).
- Glued laminated timber (Glulam).
- Building frames.
- Roof trusses (including "gang nail" trusses).
- Trussed partitions.
- Suspended floors.
- Preservative treatments.

Exclusions

- Repair of decayed timber.
- Fire protection of timber (cladding).
- Temporary works (falsework, shuttering, etc.).
- Support to suspect buildings.
- Surface painting.

Major hazards

Refer to the Introduction for details of accident types and health risks

Falls from height
Erection sequences during installation, edge protection, platforms, access.

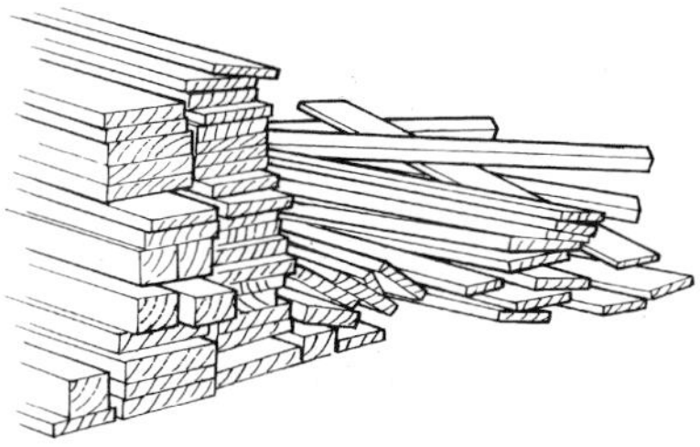

Collapse
Temporary stability. Inadequate staging/support.
Ground support for cranes.
Stacked materials.

Health hazards
Wood preservatives.
Adhesives, resins.
Treated timber.
Wood dust.

Access
Site layout, delivery routes
Storage, cranage, erection sequences.

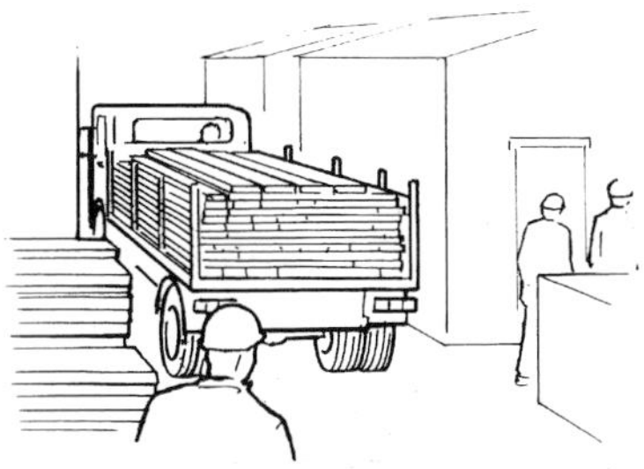

Mobile plant
Delivery lorries, site transport, cranes, dumpers, inadequate working space, erection sequences.

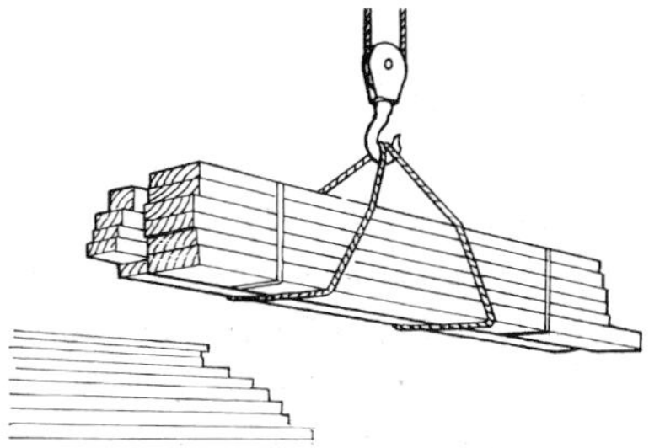

Handling
Off-loading, stacking and erecting components (manual and use of crane).
Metal connectors/cleafs.
Cuts and splinters.

C8 Group C – Primary structure
8 – Timber

Specific hazard identification

Possible key considerations:

- **Assumptions** – What are the design assumptions? Do they imply health or safety hazards?
- **Handling and storage** – How will the components be off-loaded and stored?
- **Sequences** – What are the construction sequences?
- **Access –** How will adequate access and working space be maintained?
- **Weather** – What are the likely weather conditions?
- **Installation** – What procedures are necessary for unit installation?
- **Temporary support –** What provisions will be necessary?
- **Fire walls** – What is construction against fire walls?
- **Preservatives** – What chemicals will be applied?
- **Any others?**

Prompts

- Temporary conditions
- Transport
- Access
- Lifting points
- Positioning, alignment
- Connections
- Tolerances
- Preservatives
- Adhesives
- Demolition

Hazards consideration in design

Stage	Considerations and issues	Possible design options to avoid or mitigate hazards identified for timber
Concept design	• Why use timber? • Scope of work • Interfaces • Site environment • Programme • Outline sequences	• The alternatives to timber will usually include steel or concrete. • Fire prevention is a major consideration when using timber. • Study the options for access and working space, in conjunction with construction sequences available within the constraints of the site. • Examine the feasibility of just-in-time delivery of pre-fabricated units.
Scheme design	• Major equipment • Access • Component sizes and weights • Working space • Storage areas • Handling methods • Installation • Pre-assembly • Cranage	• Review component sizes and weights together with handling methods. • Develop arrangements for just-in-time delivery, to enable direct placement of the unit into the final position. • Identify handling methods to suit methods of supply and storage. • Examine sequences of construction, to avoid unnecessary overlapping of activities and other trades.
Detail design	• Temporary supports • Preservatives • Tolerances • Connections • Lifting points • Positioning	• If mechanical handling is necessary, incorporate lifting and handling measures into the design. • Devise permanent works to incorporate temporary support. • Develop joint arrangements which minimise cutting on site. • Specify preservatives that are not hazardous to operatives during application; preferably apply preservatives in factory conditions.

Examples of risk mitigation (methods of solution)

ACTION / ISSUE	Roof erection
Avoidance Design to avoid identified hazards but beware of introducing others	Design for pre-fabricated roof at ground level, place by crane.
Reduction Design to reduce identified hazards but beware of increasing others	Lift trusses by crane and provide space for use of hydraulic platforms.
Control Design to provide acceptable safeguards for all remaining identified hazards	Incorporate handling points and positive seating.

Examples of risk mitigation (issues addressed at different stages)

A leisure facility is to include a large enclosed space. The architect wishes to incorporate a timber finish to the main structural members. Site investigation results indicates that a lightweight structure is required to provide acceptable bearing pressures.

Concept design	Scheme design	Detail design
Structural members are to be designed using glued-laminated timber columns and timber roof trusses. Use of timber means lower capacity cranes for erection. These can be used working within the building on a prepared granular base.	Specify allowable tolerances and temporary wind bracing requirements to ensure stability during erection. Lightweight tiles to be used.	Design pre-fabricated crane-handled hips and valley sets to minimise working at height.

A timber structure is to be constructed against an existing Grade I Listed Building. Due to the unconformity of the building, significant site cutting and jointing is envisaged. The cut surfaces will need to be treated with preservatives.

Concept design	Scheme design	Detail design
Design members to maximise pre-fabrication and minimise make-up pieces and site application of preservatives.	Specify application requirements for preservative treatment. Maximum in pre-fabrication works by vacuum pressure treatment. Design make-up pieces.	Design site connections to make application of preservatives easily accessible and minimise working at height.

Related issues

References within this document

References	Related issues
A1	Surrounding environment will need to be considered when examining access to the installation locations.
A2	Site clearance and simple demolition may be relevant in providing adequate access and storage areas.
A3	Site investigation will usually need to be carried out to determine criteria for the design of the timber structure.
A4	Access onto site and adjacent to installation location will need to be considered to ensure safe construction.
A5	Site layout may interact with decision to use timber construction.
D2	Roofs are likely to be constructed using timber trusses.
E6	Working over/near water may be an important consideration when working on timber construction.

References for further guidance

Primary general references and background information given in Section 4

Primary

- TRADA Timber in construction
- TRADA Knowledge Centre at www.trada.co.uk
- CIRIA R111 (updated 1994) Structural renovation of traditional buildings
- CIRIA SP57 (1988) Handling of materials on site

Secondary

- Ozelton, E C and Baird, J A, Timber Designer's Manual (Blackwell)
- Gang-Nail Systems (1989) The trussed rafter manual (Gang-Nail Systems)
- BS 5268: Part 5:1989 Code of practice for preservative treatment of structural timber

Background

- BS 5268: Part 3:1985 Code of practice for trussed rafter roofs

Classification

CAWS Group G, **CEWS** Class O, **CI/SfB** Code (2-)i

Group D – Building elements and building services
1 – External cladding

D1

Read the Introduction before using the following guidelines

Scope

- Factory fabricated panel cladding systems.
- Site assembled traditional frame systems.
- Panel rainscreen cladding systems.
- Panel overcladding systems.
- Thin stone and brick veneers.
- Profiled metal sheeting including composite panels.

Exclusions

- Brickwork/blockwork (see C7).
- Insulated and non-insulated renders.
- Special glass assemblies (eg Pilkington's Planar Glazing)
- Special finishes (eg patination).
- Restoration.

Major hazards

Refer to the Introduction for details of accident types and health risks

Working at height
Workers falling during erection, cleaning, maintenance and repair. Dropped objects during erection, maintenance and repair.

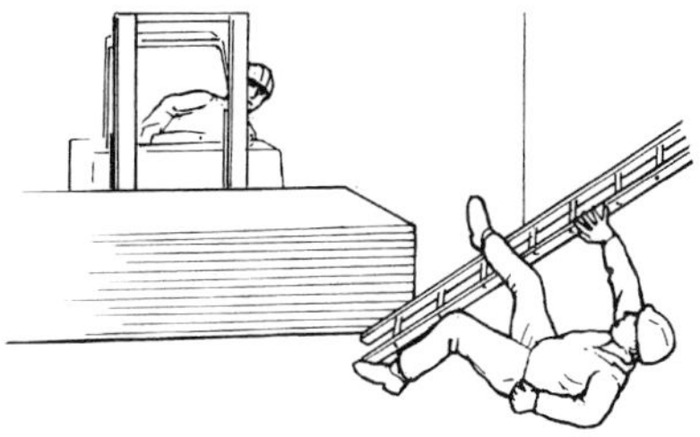

Replacement
Access, size of pieces to be moved round building and into position, ease of removal, safe fixing.

Glazing
Accidental breakage, site glazing, replacement glazing from outside, large pieces of glass, annealed glass.

Handling/transportation
Many stages - on/off lorries, in/out of storage, up/down building, into final position on building, repair/maintenance.

Hazardous operations
Surveys prior to installation. Glazing, testing, site welding, cutting, maintenance.

Hazardous substance
Paints, sealants, insulating materials.

Group D – Building elements and building services
1 – External cladding

Specific hazard identification

Possible key considerations:

- **Building location** - Are there particular access problems? Do the public have to be protected? Will the site exclude using particular plant, cladding system or materials?

  Building form - Can the cladding be prefabricated? Can large panels be installed and fixed safely to all elevations? Are there any difficult to construct features (eg overhangs)?
- **Repair and maintenance** - What suitable methods and equipment for cleaning, inspection and maintenance? Will this equipment allow replacement of pieces of cladding (eg glass)?
- **Life of cladding** - Will some or all elements of the cladding need replacement? Can replacement be carried out safely? Can the cladding be dismantled safely?

Prompts

- Materials (heavy/light)
- Size of pieces (man lift or machine)
- Life to first maintenance and failure maintenance
- Glaze from inside or out
- Replacement
- Fixings (easy to install)
- Access to site
- Access round the building
- Adjustment/alignment/tolerances
- Any restricted spaces
- Jointing/sealing
- Cleaning
- Large and lightweight panels
- Wind loading effect

Hazards consideration in design

Stage	Considerations and issues	Possible design options to avoid or mitigate hazards identified for external cladding
Concept design	• Location • Building shape • Form of construction • Access for plant/people • Cleaning, repair and maintenance • Sequence of operations • Method of assembly • Element size/weight • Interfaces • Temporary conditions • Connections/fixings • Adjustment/alignment • Testing	• Study forms of cladding construction that allow safe working space for operatives and safe lining and levelling once on the building. • Check whether planning/budgetary/client restraints permit the use of permanently installed facade cleaning and maintenance equipment. • Identify variations to the building shape which suit cladding that can be constructed, cleaned and maintained safely.
Scheme design		• Develop cladding details to provide primary attachments that are safe and easy to use in all locations. • Develop cladding jointing details that permit safe and easy installation. • Design elements so that they can be cleaned and replaced safely.
Detail design		• If mechanical handling is required, incorporate lifting and handling measures into the design. • If restraints are required for cleaning and maintenance equipment incorporate these into the design. • Co-ordinate with designer of structural frame to ensure primary fixings are safe and easy to install.

Examples of risk mitigation (methods of solution)

ACTION \ ISSUE	Falls during cleaning	Hazardous operation fixings
Avoidance Design to avoid identified hazards but beware of introducing others	Provide permanently installed access equipment that can be entered without stepping over a building edge.	Provide primary attachments to the frame that slot together without the need for loose bolts.
Reduction Design to reduce identified hazards but beware of increasing others	Provide suitable access round the building to allow lorry-mounted access platforms to reach all parts of the facade.	Provide primary attachments to the frame that allow bolts to be placed without the need to work overhead.
Control Design to provide acceptable safeguards for all remaining identified hazards	Provide a comprehensive system of attachments and latchway wires for personal protective equipment.	Provide a greater number of primary attachments to the frame so that smaller bolts can be used.

Examples of risk mitigation (issues addressed at different stages)

An existing office building in the centre of a major city is to be refurbished and reclad. The building is bounded on two sides by existing buildings and on the other two sides by very busy city streets and public pavements. There is no room for a crane or a large hoist, and the internal goods lift is small.

Concept design	Scheme design	Detail design
Large prefabricated panels cannot be lifted. A traditional aluminium framing system is appropriate with infill elements designed to fit the lifting arrangements and access round the building.	The sequence of operations is planned so that the elements needed to complete sections of wall are delivered to site in crates of a size to suit the lifting arrangements and access round the building.	Design the primary attachments so that they can be safely installed from the inside at each floor level.

The project involves the overcladding of a seven-storey block of local authority flats. The block is brick-clad, concrete framed, with timber-framed single-glazed windows. The local authority wants to change the municipal appearance, improve thermal performance and reduce maintenance costs without moving the occupants.

Concept design	Scheme design	Detail design
A metal rainscreen overcladding system with insulation behind meets the brief. The issues associated with fixing insulation and the problems of noise/dust/isolation when installing fixings need to be considered.	A cleaning and maintenance strategy is developed to allow windows to be cleaned from inside and the main facade from roof mounted access equipment.	A sequence of operation is developed for removing existing windows and filling new ones into the overcladding system without too much disruptions to the occupiers.

Group D – Building elements and building services
1 – External cladding

Related issues

References within this document

References	Related issues
D6	Cleaning and maintenance is fundamental to cladding design and must be considered at the earliest possible time.
A4	Access onto and around the site may dictate the cladding that has to be used.
C1-C8	Primary structure will dictate the type of fixing and the amount of adjustment required to take up tolerances safely.
	Interfaces for different trades will dictate sequencing of operations and replacement and maintenance strategies.
	Operators should be aware that maintenance and repair companies are sometimes less knowledgeable than the original installers/designers. Keep maintenance and repair to a minimum by good design and the use of appropriate materials.
	Occupied buildings, maintain safe access/egress and fire escape if the building is occupied while work is in progress.

References for further guidance

Primary general references and background information given in Section 4

Primary

- CIRIA SP57 (1988) Handling of materials on site
- Glass and Glazing Federation – Working with glass
- BS 6262:1982 Code of practice for glazing for buildings
- BS 6037:2003 Code of practice for the planning, design and installation and use of permanently installed access equipment. Suspended access equipment
- BS 8213:1991 Part 1 Code of practice for safety in use and during cleaning of windows and doors

Secondary

- Standing Committee on Structural Safety (SCOSS)
  - Report No 8 (1989) – Section 3.7 Cladding failures
  - Report No 9 (1992) – Section 3.7 Cladding and controls over cladding repair and refurbishment
  - Report No 10 (1994) – Section 3.3 Special structures, components and materials, sub-sections cladding and glazing
- HSE CIS5 (revised) Temporarily suspended access cradles and platforms
- CIRIA B5 (1988) Rainscreen cladding, a guide to design principles and practice
- CIRIA C524 (2000) Cladding fixings

Background

- BS 6206:1981 (1994) Specification for impact performance requirements for flat safety glass and safety plastics for use in buildings
- BS 2830:1994 Specification for suspended access equipment for use in the building, engineering, construction, steeplejack and cleaning industries
- BS 5974:1990 Code of practice for temporarily installed suspended scaffolds and access equipment
- HSE PM30 Suspended access equipment

Classification

CAWS Group H, **CEWS** Class Z, **CI/SfB** Code (41)

Group D – Building elements and building services
2 – Roof coverings and finishes

Read the Introduction before using the following guidelines

Scope

- Flat roofs.
- Pitched roofs.
- Terraces and balconies.
- Car park/service decks.
- Skylights and rooflights.

Exclusions

- Fabric and membrane roofs (tents).
- Temporary protection roofs.
- Mansard roofs.
- Prefabricated domes, etc.

Major hazards

Refer to the Introduction for details of accident types and health risks

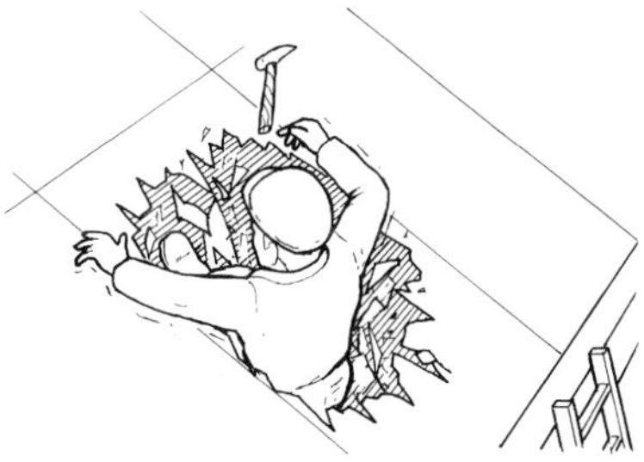

Falls from heights
Working at heights on steep pitches and slippery surfaces. Fragile roofs/rooflights. Lack of edge protection.

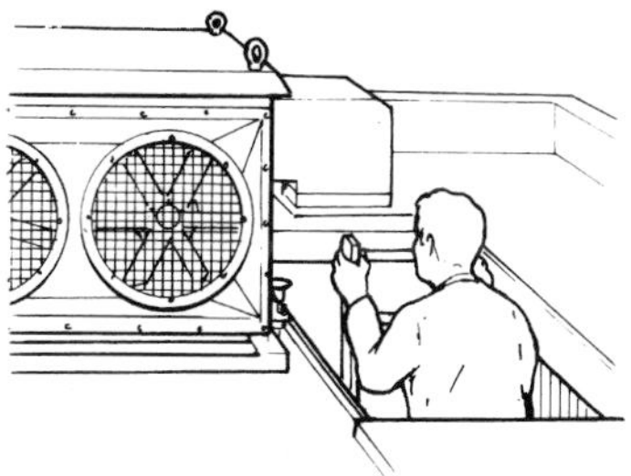

Access
Maintenance of roof and plant. Permanent or occasional use by building occupants, contractor (eg fire escape).

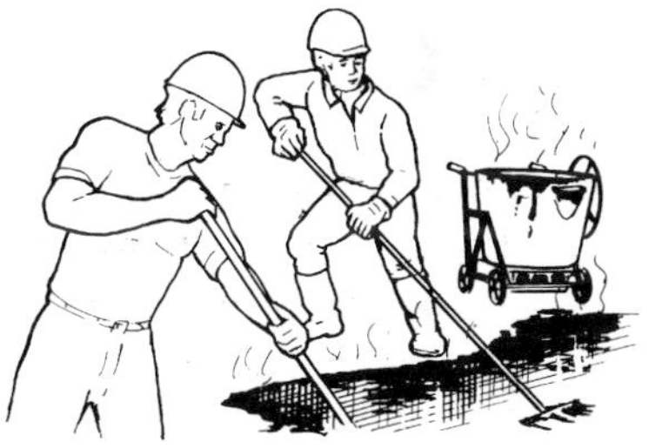

Hazardous materials and Hazardous operations:
Hot asphalt and bitumen, solvents and volatiles. Welding and soldering of metalwork.

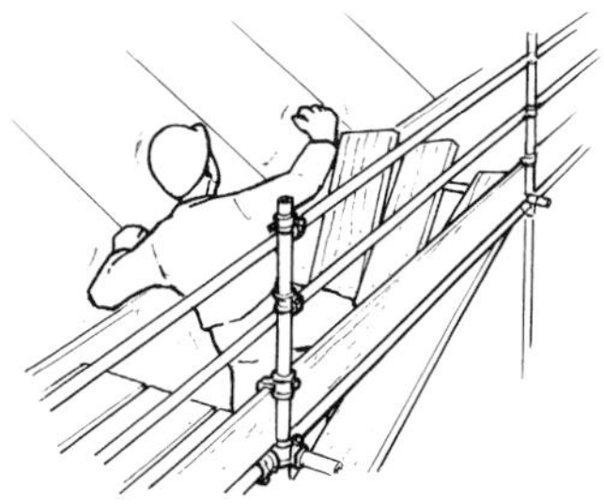

Collapse
Temporary support structure, construction platform and permanent works.

Handling
Transport and handling of materials to and across the roof. Hazards from weight, size, temperature or chemical nature of materials.

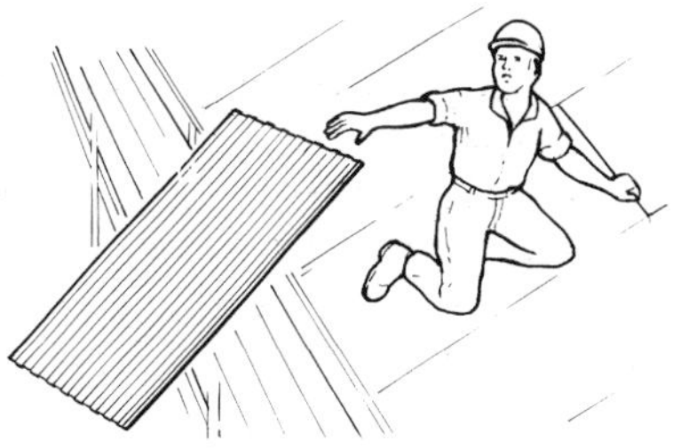

Falling or dropping
Onto people below roof level. Items falling from unprotected roof perimeter or fragile zones within roof area (eg rooflights).

Group D – Building elements and building services 2 – Roof coverings and finishes

Specific hazard identification

Possible key considerations:

- What are the benefits of pitched versus flat roofs?
- What provisions and assumptions are made for temporary stability and safety?
- What provision is made for the effects of wind during construction and maintenance?
- Is the roof construction designed to minimise risks from exposure to chemicals (solvents/fumes) and hot materials?
- What are the weather conditions likely to be during installation?
- What are the fire risks?
- What people access is proposed for roof maintenance?
- Are there general or localised weak zones in the accessible roof area (eg. rooflights)?
- Can permanent edge protection/protected walkways be incorporated?
- Size and weight vs handling and wind resistance.

Prompts

- Stability and support
- Access
- Transport of materials
- Solvents and volatiles
- Hot working
- Rainwater disposal
- Windspeed and exposure
- Falling/dropping
- Design life
- Maintenance
- Edge protection
- Panel/roll size
- Fragility of materials
- Flashings (lead vs EPDM)

Hazards consideration in design

Stage	Considerations and issues	Possible design options to avoid or mitigate hazards identified for roof coverings and finishes
Concept design	• Flat versus pitched • Building structure influence on roof type and covering. • Wind/rain exposure • Sequence of construction activities • Access to site and roof • Stability and support • Selection of roof system • Maintenance access • Lead flashing	• A flat roof can provide a more secure work platform and access for maintenance. • A pitched roof can have a longer design life and require less maintenance.
Scheme design		• Security of roof components during construction and service will be affected by wind speed. • Particular wind phenomena occur at near roof perimeters and ridge. • The sequencing of roof construction will influence the risks of falling, temporary instability, weather hazards and problems of temporary storage on the roof. • Consideration needs to be given to the stability and support of each stage - temporary or permanent - of the roof construction; allow for possible loads at both stages.
Detail design		• Selection of materials, handling and fixing to minimise hazards from installation. • When the design options will determine a sequence of working, favour the option/sequence which minimises the risks from temporary instability, lack of support, inclement weather working, difficult local access - partial covering of work by other traders.

Examples of risk mitigation (methods of solution)

ACTION \ ISSUE	Flat roof membrane	Translucent fragile panels
Avoidance Design to avoid identified hazards but beware of introducing others	Loose lay cold membranes to avoid the use of hot bitumen.	Use alternative stronger material.
Reduction Design to reduce identified hazards but beware of increasing others	Use torch-applied bonding or solvents/adhesives to secure pre-formed products.	Use a thicker material or additional layers of material.
Control Design to provide acceptable safeguards for all remaining identified hazards	Specify safety provisions.	Provide a walkway/grille over the fragile panels.

Examples of risk mitigation (issues addressed at different stages)

A large factory requires fast construction to bring the facility on stream. It is preferred that the roof be finished early in the programme so that other construction and fitting out can take place beneath in protected conditions.

Concept design	Scheme design	Detail design
Cost, programme considerations and repetitive nature of construction lead to choice of flat roof which can quickly follow frame erection. Rapid placement of metal decking not only provides safer work platform for use at height but also protects remainder of workforce at ground level from dropped objects and inclement weather.	Roofing materials to be mechanically fixed to minimise time spent at height and complexity/hazards of these work activities.	Final detailing of gutters/eaves to incorporate provision for edge protection primarily for construction, but which can also be exploited during maintenance.

Design of a pitched roofed warehouse with valley gutters.
The "traditional" roof appearance is preferred because of existing architecture and technical concerns.

Concept design	Scheme design	Detail design
Consider choice of roofing sheets. Reject fragile material as roof access is required for cleaning gutters.	Appropriate non-fragile material selected.	Make valley gutter wide enough and strong enough to walk along. Provide permanent edge protection at gable ends of gutters.

Group D – Building elements and building services
2 – Roof coverings and finishes

A GENERAL PLANNING
B EXCAVATIONS AND FOUNDATIONS
C PRIMARY STRUCTURE
D 2 BUILDING ELEMENTS AND BUILDING SERVICES
E CIVIL ENGINEERING

Related issues

References within this document

References	Related issues
A1	Surrounding environment will need to be assessed for access routes for materials and prevailing wind conditions.
A4	Access onto and within the site will need to be considered.
A5	Site layout could be influenced by roof construction type and associated methods.
C1-C8	Primary structure can influence roof construction options.
C8	Timber will often be used for trusses supporting roof coverings.
D1	External cladding details and roofs interact.
D4	Windows/glazing may be included in roofs and will require cleaning.
D7	Mechanical plant for services often have to be roof-mounted.
D8	Lightning protection.

References for further guidance

Primary general references and background information given in Section 4

Primary

- HSG33 Health and safety in roofwork
- CITB – Construction site safety, Vol 1, Section C6, Safe working on roofs and at heights (2003)
- BEC Construction Safety, Section 15 Work on roofs

Secondary

- BS 6229:2003 Code of practice for flat roofs with continuously supported coverings
- BS 5534: 990 Code of practice for slating and tiling, Part 1
- BS 6915:2001 Code of practice for design and construction of fully supported lead sheet roof
- HSE CIS5 (revised) Temporarily suspended access cradles and platforms
- HSE L132 – Approved code of practice and guidance – Control of lead at work regulations 2002

Background

- Copper in Roofing Pocket Book
- BS 6399 Loading for buildings: Part 2 (1997) Wind loads
- BS 6399: Part 2: 1995 Code of practice for wind loads
- CIRIA and BFRC B15 (1993) Flat Roofing guide: Design and good practice
- BRE Digest 346: Assessment of wind loads
- BRE Report 302 (1996) Roofs and roofing: performance, diagnosis, maintenance, repair and the avoidance of defects (BRE)

Classification

CAWS Group H, **CEWS** Class W, **CI/SfB** Code (47)

Group D – Building elements and building services
3 – Atria

D3

Read the Introduction before using the following guidelines

Scope

- Walls to atria.
- Roofs to atria.
- Roofs to shopping malls.

Exclusions

- Internal partitions.
- Rooflights and skylights.

Major hazards

Refer to the Introduction for details of accident types and health risks

Working at height
Falling from height,cleaning, maintenance and repair, access.
Dropped objects.

Hazardous materials
Paints, sealants, glass.

Handling
Many stages on/off lorries, in/out of storage, up/down building into final position on building, repair maintenance.

Glazing
Site glazing, glazing from outside, large pieces of glass, replacement, accidental breakage, metal/glass contact.

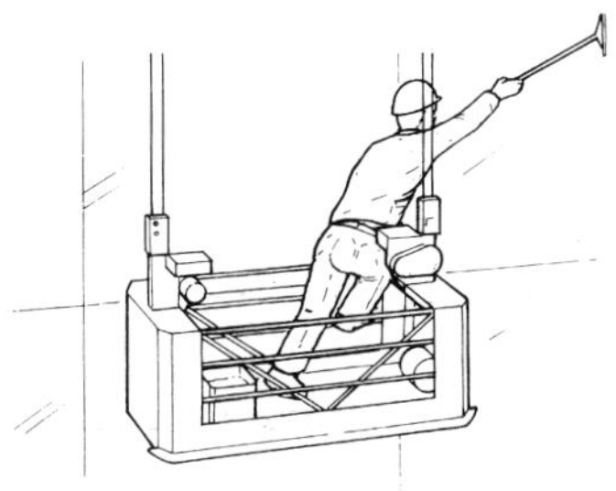

Hazardous operations
Inspections, glazing and reglazing, cleaning and maintenance, servicing smoke vents.

D3 Group D – Building elements and building services 3 – Atria

Specific hazard identification

Possible key considerations:

- **Installation** - Has access for installation been considered? Do the public have to be protected? Will scaffolding be required? How are elements to be lifted? Fixed?
- **Cleaning and maintenance** - Consideration to be given to cleaning inside and outside. How is access for cleaning and maintenance to be achieved? Consider how replacement is carried out. Will other services interfere with cleaning?
- **Injury from falling objects** - Consider the use of safe materials overhead (eg laminated glass)? Can features be designed to protect people below from falling objects? Consider the use of handrails and other barriers to stop people and objects striking glass.

Prompts

- Access
- Maintenance
- Replacement
- People overhead
- Handling
- Occupied premises
- Tolerances
- Fixings (easy to reach)
- Jointing and sealing
- Other services
- Confined spaces

Hazards consideration in design

Stage	Considerations and issues	Possible design options to avoid or mitigate hazards identified for atria
Concept design	• Access • Cleaning and maintenance • Form of construction • Shape and size • Design life • Element size/weight • Sequence • Cranage • Repair replacement • Types of glass and configuration • Connections/fixings • Assembly • Temporary conditions during erection • Adjustment/ • Alignment • Surveys of supporting structure • Testing of cleaning equipment • Sequence of assembly	• Consider using purpose-designed permanently installed access equipment inside and outside overhead glazing for cleaning, maintenance and glass replacement. • Ensure safe access can be provided for installation • Ensure access to cleaning and maintenance equipment provided can be achieved safely.
Scheme design		• The design of the glazing system and associated cleaning equipment should maximise factory prefabrication. • Any external temporary scaffolding roofs required for construction should be designed to allow smoke extraction in the event of a fire.
Detail design		• Fixing details for overhead glazing need to be detailed so that they can be secured without working overhead and blind. • Glazing systems need to be designed so that glass can be replaced without danger of whole sheets falling onto operatives.

Examples of risk mitigation (methods of solution)

ACTION \ ISSUE	Falls during maintenance	Injury from falling objects
Avoidance Design to avoid identified hazards but beware of introducing others	Provide permanently-installed access equipment inside and outside the roofs to atria that can be entered without stepping over a building edge or onto the roof.	Use materials that will not fall out of supporting framework if they are broken accidentally.
Reduction Design to reduce identified hazards but beware of increasing others	Consider a strategy for maintenance that involves the regular erection of purpose designed scaffolding with access platforms at appropriate levels to allow safe working methods to be used.	Provide features below that will catch falling objects and direct them to areas where they will do no harm.
Control Design to provide acceptable safeguards for all remaining identified hazards	Provide a comprehensive system of attachments and latchway wires for personal protective equipment, with safe access onto these systems.	Prevent or limit access to the vulnerable elements by barriers or protection.

Examples of risk mitigation (issues addressed at different stages)

A glass roof over an atrium has suffered from spontaneous fracture of toughened glass panes due to nickel sulphide inclusions. The owner wishes to re-glaze the roof while the building is still in use. The base of the atrium is 5 storeys below the roof and is used as a circulation area.

Concept design	Scheme design	Detail design
A scaffolding will be erected with a working deck just below the roof. Temporary roofing will go over the glass roof.	Consider options for supporting the working deck/crash deck and temporary smoke control.	Identify risks associated with the assumed sequence and method of dismantling.

A new shopping centre is to be constructed. The mall between the shops is to have a glazed roof.

Concept design	Scheme design	Detail design
Consider various forms for the glazed roof that will allow for cleaning and maintenance. Assess options for access to building and maintenance equipment	Consider various types of cleaning equipment that can be used and select the one that offers the easiest and safest access for installation and maintenance.	Design safe and unobstructed access for maintenance cleaning equipment.

Related issues

References within this document

References	Related issues
A4	Access onto the site and access for installing and cleaning atria cladding will dictate the form of cladding to be used and the sequence and method of assembly.
C1-C8	Primary structure will dictate the types and method of fixing and the amount of adjustment required to take up tolerances.
D6	Cleaning and maintenance is fundamental to atria design and must be considered at the earliest possible time.
	Interfaces for different trades will dictate sequencing of operations and replacement and maintenance strategies.
	Operators should be aware that maintenance and repair companies are sometimes less knowledgeable than the original installers/designers. Hence minimise maintenance and repair in design and use appropriate materials.
	Occupied building, maintain safe access/egress and fire escape if the building is occupied while construction cleaning or maintenance work is in progress.

References for further guidance

Primary general references and background information given in Section 4

Primary

- Glass and Glazing Federation. Code of Practice for window installation safety
- Glass and Glazing Federation. Working with glass
- BS 5516:1991 Code of practice for design and installation of sloping and vertical patent glazing
- CIRIA Publication SP87 (1992) Wall technology volumes A to G

Secondary

- Standing Committee on Structural Safety (SCOSS)
  - Report No. 8 – Section 3.7 Cladding failures
  - Report No. 9 – Section 3.7 Cladding & Controls over cladding repair and refurbishment
  - Report No. 10 – Section 3.3 Special structures, components and materials, sub-sections cladding and glazing
- BS 6037:2003 Code of practice for permanently installed suspended access equipment.
- Dawson S, Cleaning System for Glass Dome. Architects Journal, Jan 98

Background

- CIRIA SP57 (1988) Handling of materials on site
- HSE CIS5 (revised) Temporarily suspended access cradles and platforms
- HSE PM30 Suspended access equipment
- BRE Report 258 (1994) Design approaches for smoke control in atrium buildings (BRE)

Classification

CAWS Group H, **CEWS** Class Z, **CI/SfB** Code (--) 913

Group D – Building elements and building services
4 – Windows/glazing including window cleaning

Read the Introduction before using the following guidelines

Scope

- New windows.
- Replacement windows.
- Windows in curtain walling.
- Windows in overcladding.
- Windows in brickwork/blockwork/stonework.
- Domestic windows.

Exclusions

- Rooflights/skylights (see D2).
- Special loading conditions (eg bomb blast).
- Smoke ventilators.

Major hazards

Refer to the Introduction for details of accident types and health risks

Working at height
Falling from height, cleaning, maintenance and repair, access, dropped objects.

Cleaning and maintenance
Outside of window difficult/dangerous to reach, rarely provision for access and working place.

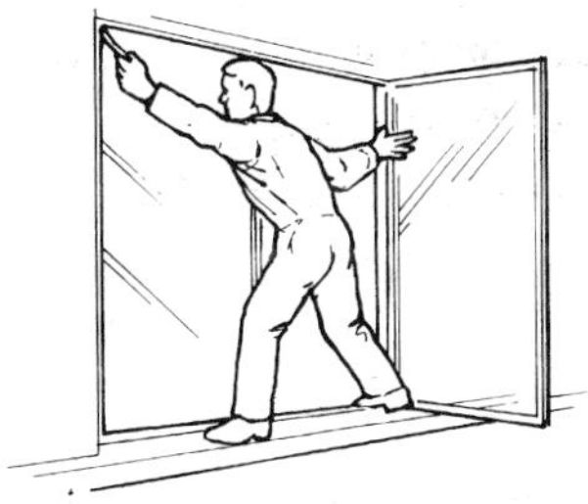

Replacement
Access through occupied spaces and lack of protection outside.

Handling transportation
Many stages, on/off lorries, in/out of storage, up/down building, into final position on building.

Hazardous operations
Surveys prior to installation, glazing, testing cleaning and maintenance.

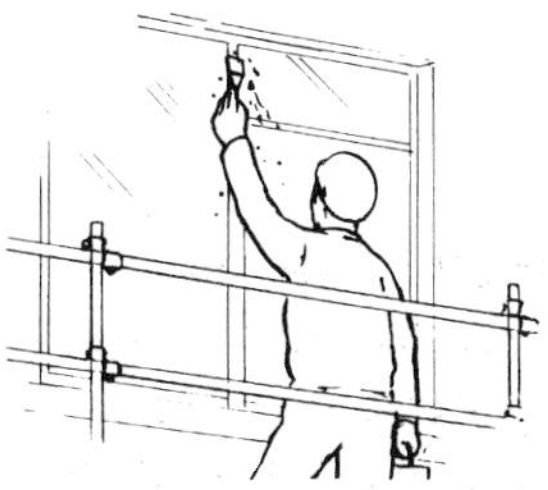

Hazardous materials
Paints, sealants, insulating materials, glass.

D4 Group D – Building elements and building services
4 – Windows/glazing including window cleaning

Specific hazard identification

Possible key considerations:

- **Installation** - Has access for installation been considered? Will the site exclude using particular plant or equipment? How are windows lifted? How are they fixed? Can the pieces be easily handled, moved and installed?
- **Cleaning and maintenance** - How is access for cleaning and maintenance achieved? How is replacement carried out? How will windows be opened for cleaning? Will size of opening light limit cleaning and maintenance options? Will permanently installed equipment be provided for the outside of the building.
- **Occupied buildings** - will the windows be installed, cleaned and maintained from inside the building? Will special tools or procedures be required to remove windows or glass in the event of failure? Access and escape routes need to be maintained at all times. Reduce hazardous operations to a minimum. Will other services interfere with installation, cleaning and maintenance?

Prompts

- Space to align local access
- Low cills
- Safe reach to outside
- Replacement
- Building in use
- Low maintenance materials
- Glaze from inside
- Adjustment/alignment
- Size of glass pieces

Hazards consideration in design

Stage	Considerations and issues	Possible design options to avoid or mitigate hazards identified for windows
Concept design	• Building shape/size • Location • Cleaning strategy • Repair/maintenance • Form of construction • Opening type • Size of opening light • Temporary conditions • Connections fixings • Types of glass • Limiting stays • Opening devices • Locking devices • Sequence of assembly • Fixing details • Glazing details	• Consider options for cleaning maintenance and repair which include permanent or temporary access equipment. • To reduce maintenance requirements consider low maintenance options like coated aluminium or UPVC rather than timber. • Consider types of window to assist with cleaning and maintenance, eg centre pivot or reversible.
Scheme design		• Replacement strategy will dictate the size of glass and window framing that can be moved through an occupied building. • Design of fixing should consider ease of adjustment for building tolerances and easy access from inside for replacement.
Detail design		• Detail stays, restraints or guard rails to prevent window cleaners from falling out. • Detail components so that each element can be fixed and weathered without blind or overhead working.

Examples of risk mitigation (methods of solution)

ACTION \ ISSUE	Falls during cleaning
Avoidance Design to avoid identified hazards but beware of introducing others	Clean externally from permanently installed access equipment that can be entered without stepping over a building edge, and limit size of windows so that they can be cleaned internally standing on the floor.
Reduction Design to reduce identified hazards but beware of increasing others	Provide centre pivot or reversible windows of limited size so that both sides can be cleaned safely from inside the building standing on the floor.
Control Design to provide acceptable safeguards for all remaining identified hazards	Provide inward opening windows of limited size and with locking stays so that both sides can be cleaned safely from inside the building standing on the floor and provide guard rails to protect the opening while the window is open.

Examples of risk mitigation (issues addressed at different stages)

A high rise office block by the sea is to have new windows fitted. Complaints about the air-conditioning from the users require the new windows to provide natural ventilation. The building will remain occupied during the work, though there will be a phased decanting of floors to allow the recladding work to proceed.

Concept design	Scheme design	Detail design
Consider whether reversible inward opening or pivot windows in the sizes required can be safely opened for cleaning in a windy environment. Can windows of the required size be installed from inside the building or does a strategy for installation, maintenance and replacement from outside have to be developed.	Consider the various types of the window selected and develop sequences of operations for installation and replacement that offer the easiest and safest means of access and egress.	Specify a window type that allows only narrow opening in everyday use, but wider opening with a special key for cleaning.

An extension of an existing building requires windows to match existing Georgian sash windows, but for acoustic reasons need to be triple glazed. The stone facade will not be cleaned regularly, the windows need to be cleaned regularly.

Concept design	Scheme design	Detail design
Establish a sequence of opening procedures that allows all faces of each piece of glazing to be reached from inside the building.	Confirm that the inward opening elements do not trap the cleaner or cause him to be pushed out of the window.	Detail the operating handles and safety catches to match the required sequence of opening and cleaning.

Group D – Building elements and building services
4 – Windows/glazing including window cleaning

A GENERAL PLANNING
B EXCAVATIONS AND FOUNDATIONS
C PRIMARY STRUCTURE
D 4 BUILDING ELEMENTS AND BUILDING SERVICES
E CIVIL ENGINEERING

Related issues

References within this document

References	Related issues
A4	Access - Access onto the site and access for installing and cleaning windows will dictate the type of window used, and the sequence and method of installation.
C1-C8	Primary structure - the structure surrounding the windows will dictate the type and method of fixing and the amount of adjustment required to take up tolerances safely.
D6	Cleaning and maintenance - the strategy for cleaning and maintenance is fundamental to window design and must be considered at the earliest possible time.
	Interfaces - interfacing trades will dictate sequencing of operations and replacement and maintenance strategies.
	Operators - be aware that maintenance and repair companies are sometimes less knowledgable than the original installers/designers - so keep maintenance and repair to a minimum by good design and the use of appropriate materials.
	Occupied buildings - maintain safe access/egress and fire escape if the building is occupied while construction cleaning, maintenance or repair work is in progress.

References for further guidance

Primary general references and background information given in Section 4

Primary

- Glass and Glazing Federation. Code of Practice for window installation safety
- Glass and Glazing Federation – Working with glass
- BS 8213:Part 1: 1991 Code of practice for safety in use and during cleaning of windows and doors
- BS 6262:1982 Code of practice for glazing for buildings
- Building Regulations Part N (2000). N1 Glazing – Safety in relation to impact, opening and cleaning

Secondary

- Standing Committee on Structural Safety (SCOSS)
  - Report No. 8 – Section 3.7 Cladding failures
  - Report No. 9 – Section 3.7 Cladding and controls over cladding repair and refurbishment
  - Report No. 10 – Section 3.3 Special structures, components and materials, sub-sections cladding and glazing
- BS 2830:1994 Specification for suspended access equipment
- BS 5974:1990 Code of practice for temporarily installed suspended scaffolds and access equipment
- BS 6037:2003 Code of practice for permanently installed suspended access equipment
- HSE CIS5 (revised) Temporarily suspended access cradles and platforms
- HSE PM30 Suspended access equipment.
- Building Control Journal – Cleaning Windows 1997 (Alternative ways)

Background

- BS 6206:1981 (1994) Specification for impact performance requirements for flat safety glass and safety plastics for use in buildings

Classification

CAWS Group L, **CEWS** Class Z, **CI/SfB** Code (31.4)

Read the Introduction before using the following guidelines

Scope

- Decorative coatings (paints) for timber, steel, concrete, plaster, masonry.
- Plaster, render and screeds.
- Corrosion and fire protection coatings (dry and wet systems).
- Floor and wall tiling.

Exclusions

- Composite sheet or board linings and cladding.

Major hazards

Refer to the Introduction for details of accident types and health risks

Toxic solvents
Solvents/volatiles from treatments or cleaning materials released into work space. Toxicity/suffocation if insufficient ventilation.

Fire
Combustibility and surface fire spread (including smoke).

Access and workplatform
Safety during application and maintenance.

Hazardous substances
Manmade mineral fibres, solvents, dust (silica).

Manual handling
Transport, storage and mixing of bulky/heavy materials.

Health
Noise, fume, dust.
Resins and glues.

Group D – Building elements and building services
5 – Surface coatings and finishings

Specific hazard identification

Possible key considerations:

- **Handling and storage** - How will materials be off-loaded and stored?
- **Mixing** - What, if any, are the requirements?
- **Application method** - What are the options and implications?
- **Surface pre-treatment** - What operations are required?
- **Application** - What access to place of work is needed?
- **Weather conditions** - What are the application limitations?
- **Maintenance** - What are the requirements?
- **Materials** - What is the chemical nature?
- **Fire** - Is there a risk during application?
- **Any others?**

Prompts

- Solvents
- Transport
- Mixing
- Dust/fumes
- Maintenance
- Application equipment
- Access/place of work
- Working at height

Hazards consideration in design

<table>
<tr><th>Stage</th><th>Considerations and issues</th><th>Possible design options to avoid or mitigate hazards identified for surface coatings and finishings</th></tr>
<tr><td>Concept design</td><td rowspan="3">• Why apply a finish?
• Why spray apply finish?
• Outline sequences
• Compatibility of materials
• Working space
• Working environment
• Ventilation
• Pre-treatment
• Handling
• Access
• Materials selection
• Working sequence
• Check existing surfaces</td><td>• Can the same, or sufficient performance be achieved for the substrate without an applied finish (eg for light reflectance, or corrosion allowance)?
• To what extent does the nature of the specific material selected, and its requirements for application either reduce hazards or generate new ones?
• Sequence of construction consideration can identify situations where unacceptable hazards might occur (eg confined spaces and solvent release, working in unacceptable weather).
• Do materials or components in contact require special pretreatment which might introduce a hazard (eg chemical interaction or relative movements)?</td></tr>
<tr><td>Scheme design</td><td>• Is sufficient space allowed for the operations and maintenance?
• Is the working environment likely to be safe from hazards (eg toxic/suffocating fume release)?
• Can pre-treatment be avoided without loss of performance?</td></tr>
<tr><td>Detail design</td><td>• Choose pre-blended/pre-mixed materials to reduce handling.
• Select solvent-less or water-based products wherever possible.
• Study routes for access and handling materials and components.
• Can materials be brought safely to the point of application?
• Can the working sequence be modified to allow for better access or control of working environment?</td></tr>
</table>

Examples of risk mitigation (methods of solution)

ACTION \ ISSUE	Combustibility/fire spread	Access for application
Avoidance Design to avoid identified hazards but beware of introducing others	Use non-combustible products.	Factory preparation of substrate and application of coatings and finishes to avoid site work.
Reduction Design to reduce identified hazards but beware of increasing others	Use product of limited combustibility.	Only final layer/top coat or touching up to be site applied.
Control Design to provide acceptable safeguards for all remaining identified hazards	Limit application to low risk areas.	Access to height to be controlled by principal contractor.

Examples of risk mitigation (issues addressed at different stages)

A commercial office development wishes to offer prestige accommodation at an economic price. Space within the building for "non-productive" areas is under close scrutiny.

Plant rooms are only accessible by maintenance staff.

Concept design	Scheme design	Detail design
Plan space for optimum balance of plant requirements and maintenance of plant and surface finishes.	Provide access routes from outside building. Design adequate space for safe access to and maintenance of items of plants in plant rooms as well as access to walls, floors, etc. for maintenance of finishes.	If walls/floor need painting, select coatings which require little surface preparation, are water-based and/or solvent free and require low maintenance.

An existing warehouse is to be upgraded. The proposed works include fire - rating the steel frame.

Concept design	Scheme design	Detail design
Consider whether "dry" materials (board linings) can be used instead of "wet" materials (intumescent paint).	Consider the space available for the works and consider a method that minimises the hazards.	If "wet", specify pre-blended/mixed materials to reduce handling on site. Select solvent-free or water-based products wherever possible. If "dry", specify factory-cut boards.

Related issues

References within this document

References	Related issues
A4	Access onto site will need to accommodate frequent deliveries of materials.
A5	Site layout will need to be considered for the delivery of materials to the point of application, and the safe storage of materials.
C2	In-situ concrete often has particular finishes specified.
C5	Structural Steelwork is often site painted.
C8	Timber treatments on site need to be planned with great care.

References for further guidance

Primary general references and background information given in Section 4

Primary

- HSE CIS24 Chemical cleaners
- HSE CIS27 Solvents
- BS 5492: 1990 Code of practice for internal plastering
- BS 5262: 1991 Code of practice for external renderings
- BS 6150: 1991 Code of practice for painting of buildings
- BS 8000 Part 10: 1995 Workmanship on building sites: Code of practice for plastering and rendering
- BS 8000 Part 12: 1989 Workmanship on building sites: Code of practice for decorative wall coverings and painting
- BS 8221 Part 2: 2000 Code of Practice for cleaning and surface repair of buildings, surface repair of natural stone, brick and terracotta

Secondary

- BS 8202: Part 1 (1995), Part 2 (1992) Coatings for fire protection of building elements
- BS 8204: Part 1 (1987) to Part 5 (1994) Screeds, bases and in-situ floorings
- BS EN ISO 28028: 2000 Rubber and/or plastic hose assemblies for airless paint spraying
- CIRIA R174 (1997) New paint systems for the protection of construction steelwork

Background

- BS 6100:1991 Part 1: Section 1.3, Subsection 1.3.7, Glossary of building and civil engineering terms: Finishes (B)
- CIRIA PR78 (2000) The use of epoxy, polyester and similar reactive polymers in construction, Volume 2: Specification and use of the materials

Classification

CAWS Group M, **CEWS** Class V, **CI/SfB** Code (49)u/v

Group D – Building elements and building services
6 – Cleaning of buildings

D6

Read the Introduction before using the following guidelines

Scope

- Permanent access equipment.
- Temporary access equipment.
- Roped access.
- Hydraulic platforms (cherry picker).
- Fall/arrest systems (latchways).
- New buildings and existing buildings.
- Ladder access.

Exclusions

- Cleaning of internal spaces, eg atria (see D3).

Major hazards

Refer to the Introduction for details of accident types and health risks

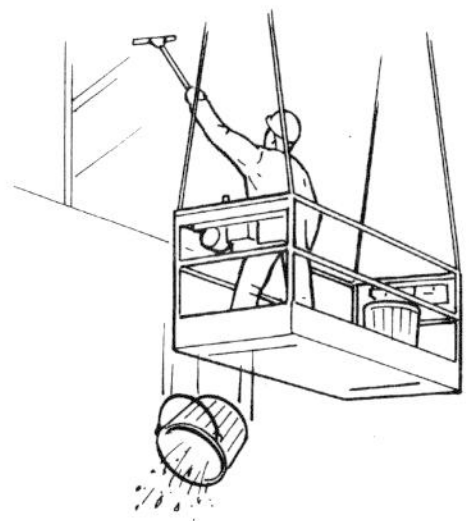

Working at height
Dropped objects and substances during cleaning. Falling out of equipment. Falling off edge of building.

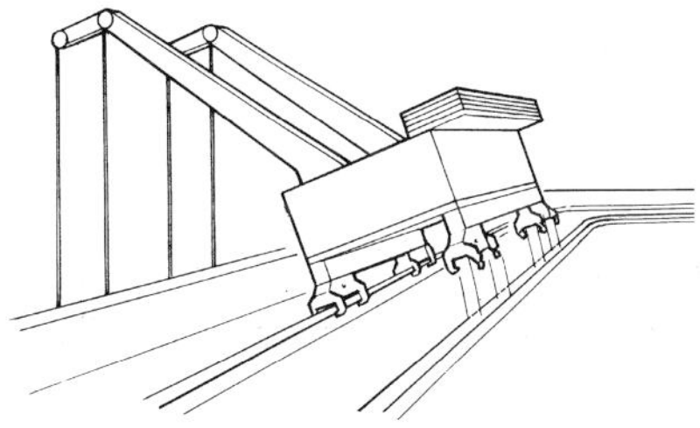

Collapse and failure
Provision of safe equipment system for cleaning.

Hazardous materials
Aggressive cleaning materials (use and removal), safe removal of waste process materials.

Weather
Ropes and equipment becoming unstable, tangled or twisted in the wind. Slippery walking surfaces.

Hazardous operations
Repair and maintenance involving cutting, heating gluing or sealing. Unsecured cleaning equipment (eg metal bucket).

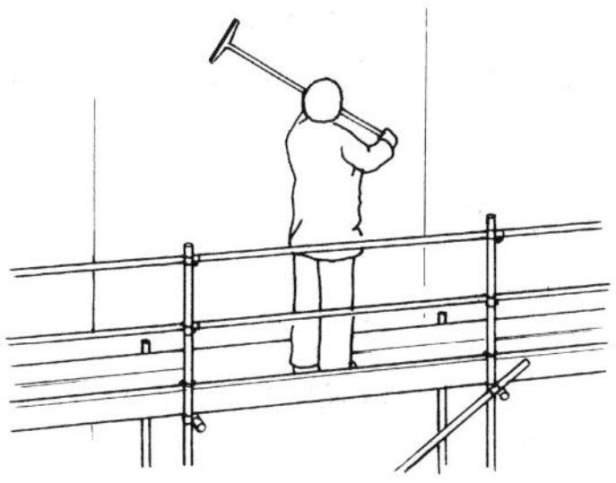

Access
Safe access and simple operation essential.

Group D – Building elements and building services
6 – Cleaning of buildings

Specific hazard identification

Possible key considerations:

- **Building location:** Are these particular access problems? Do the public have to be protected? Will the site preclude using particular equipment or methods? Is the site exposed?
- **Building form:** How is access to cleaning equipment achieved? Will cleaning operations require oversailing the public or another building? What is the form and geometry of the facade to be cleaned? Are there any overhangs to be cleaned? Is the building of exceptional size?
- **Cleaning strategy:** What, if any, specific cleaning requirements for elements of cladding? What is required frequency of cleaning? Will aggressive/harmful cleaning materials be required?
- **Does replacement and maintenance** of element need to be considered?

Prompts

- Unprotected roof edges
- Trip hazards at roof edges
- Difficult to reach areas
- Special equipment required
- Access to equipment
- Windy site
- Storage of access equipment
- Provision of warning signs

Hazards consideration in design

Stage	Considerations and issues	Possible design options to avoid or mitigate hazards identified for cleaning of buildings
Concept design	• Form of cladding • Cleaning strategy • Cleaning frequency • Materials to be cleaned • Safe working loads • Design life of cladding/finishes • Safe entry/exit • Access to all parts of the facade • Equipment to achieve access require • Review loads from equipment • Power and water supply • Attachment details • Back-up systems • Testing/commission • Service/maintenance	• Consider various types of equipment that can reach all surfaces of the building elevation without dangerous manoeuvres (eg cradle or cherry pickers). • Avoid the use of materials and finishes that require aggressive cleaning materials. • Entry/exit points for cleaning equipment need to be detailed so that dangerous manoeuvres are avoided (eg stepping over the edge of building into cradle). • Consider how cleaning may be eliminated or reduced.
Scheme design		• Develop cleaning strategy to provide full and safe coverage of the whole facade using the preferred method (eg will the stabilising legs of a cherry picker always be on solid ground?). • Develop strategy for dealing with waste products from cleaning operations.
Detail design		• With the equipment selected for cleaning, ensure that the operatives can attach personal protective equipment to secure points everywhere that cleaning and maintenance is required.

Examples of risk mitigation (methods of solution)

ACTION / ISSUE	Falls from building edge when entering cleaning cradle	Collapse of cherry picker
Avoidance Design to avoid identified hazards but beware of introducing others	Use permanently-installed roof-mounted cleaning equipment provided with luffing and slewing jib, so that access to the cradle can occur away from the roof edge.	Provide a flat, solid, clear access round the base of the building as part of the external landscaping design for the building.
Reduction Design to reduce identified hazards but beware of increasing others	Use permanently-installed roof-mounted cleaning equipment with a luffing jib, and provide safe entry points guarded by gates and rails, and with tethering points for the cradle.	Provide designated areas round the base of the building that can be accessed and which are flat, solid and clear.
Control Design to provide acceptable safeguards for all remaining identified hazards	Use permanently-installed roof-mounted cleaning equipment with a luffing jib and provide fall-arrest systems adjacent to cradle access points that can be connected/ disconnected when in the cradle.	Specify all-terrain cherry pickers with a known safety record on a variety of surfaces and sub-soil conditions. Provide clear access to the base of the building.

Examples of risk mitigation (issues addressed at different stages)

An existing office building in the centre of a major city is to be refurbished and reclad. It has a well-maintained track-mounted trolley supporting a hand- operated cradle to clean two facades and a set of fixed davit arms to support a hand-operated cradle for the other facades. The client would like to refurbish this equipment for future use.

Concept design	Scheme design	Detail design
Manually - operated equipment will be replaced by electrical operation. Review form and geometry of new cladding to establish performance requirements for new equipment.	Ensure safe access routes into the cradle.	Ensure adequate space to allow safe access for maintenance equipment.

An existing shopping centre with a glazed roof over the pedestrian malls has been recently purchased by a new landlord. The roof has been cleaned externally by operators using duck boards and ladders. The new owner wants a new purpose-designed roof access system that will allow the glass to be cleaned without anybody having to stand on the roof.

Concept design	Scheme design	Detail design
Load limitations on the existing glazing framework mean that a travelling gantry supported off the adjacent structure is required.	The travelling gantry is designed to be built adjacent to the mall glazing, and then run on temporary rails on to the permanent rails each side of the mall.	Design gantry so as to provide a safe working area.

A GENERAL PLANNING
B EXCAVATIONS AND FOUNDATIONS
C PRIMARY STRUCTURE
D6 BUILDING ELEMENTS AND BUILDING SERVICES
E CIVIL ENGINEERING

Related issues

References within this document

References	Related issues
A1	Surrounding environment may dictate operation times and frequency of cleaning.
A4	Access to site will dictate how machinery is put up and commissioned.
A4	Access to roof areas will dictate sequencing and protection.
A4,D2	Primary structure will dictate type of fixings, type and span of track, weight of access equipment.
D1-D5	Maintenance, repair and replacement strategy of facade/envelope elements needs to be considered with cleaning.
	External materials will dictate methods and frequency of cleaning.

References for further guidance

Primary general references and background information given in Section 4

Primary

- BS 5974:1990 Code of practice for temporarily installed suspended scaffolds and access equipment
- BS 6037:2003 Code of practice for permanently installed suspended access equipment
- BS 8213:1991 Part 1 Code of practice for safety in use and during cleaning of windows and doors
- HSE CIS5 (revised) Temporarily suspended access cradles and platforms
- HSE CIS24 Chemical cleaners

Secondary

- BS 2830:1994 Specification for suspended access equipment
- HSE PM30 Suspended access equipment
- CIRIA SP71 (1989) Graffiti removal and control

Background

- BS 6570: 986 Code of practice for the selection, care and maintenance of steel wire ropes
- Ashurst, N (1994) Cleaning historic buildings. 2 vols (Donhead)

Classification

CAWS Group A, **CEWS** Class A, **CI/SfB** Code (--)(W2)

Read the Introduction before using the following guidelines

Scope

- Boiler plant.
- Refrigeration and chiller plant including heat rejection equipment.
- Water treatment plant.
- Air-handling plant.
- Fans and pumps.
- Distribution systems - pipework and ductwork.
- Room units - radiators, fan coils, VAV boxes, etc.
- Gas installations.
- Steam and high temperature hot water installations.
- Smoke ventilation.

Exclusions

- Specialist containment systems - microbiological cabinets, fume cupboards.
- Systems handling hazardous radiological or biological substances.
- Control and control panels (see D8).

Major hazards

Refer to the Introduction for details of accident types and health risks

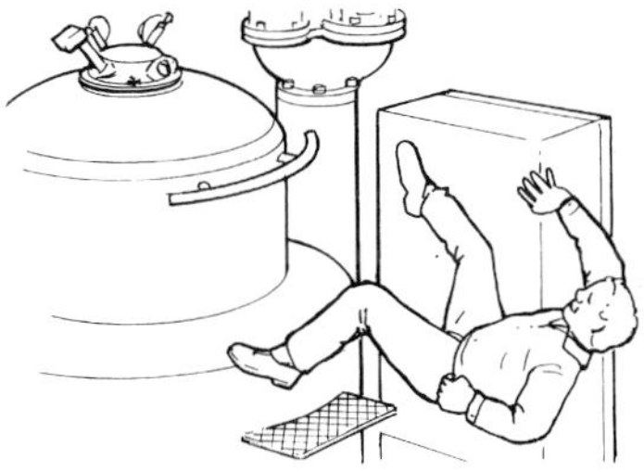

Falling from heights
Installation, commissioning and maintenance of equipment.
Builders work openings

Manual handling
Mechanical equipment is generally large and heavy, or simply bulky.

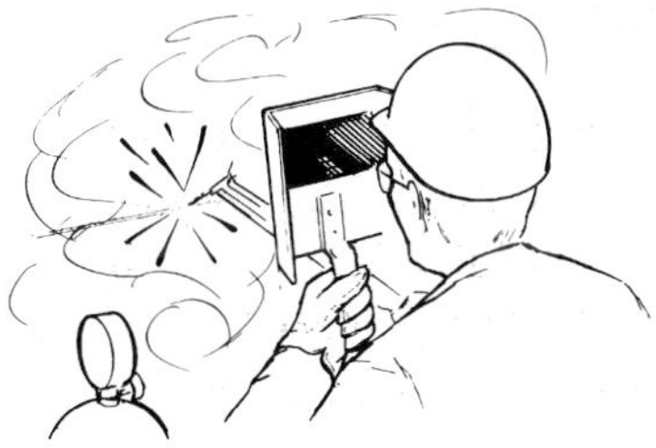

Hazardous atmospheres/substances
Explosive gas, dusts or vapours, welding fumes, products of combustion, fine fibres, refrigerants, legionella, etc.

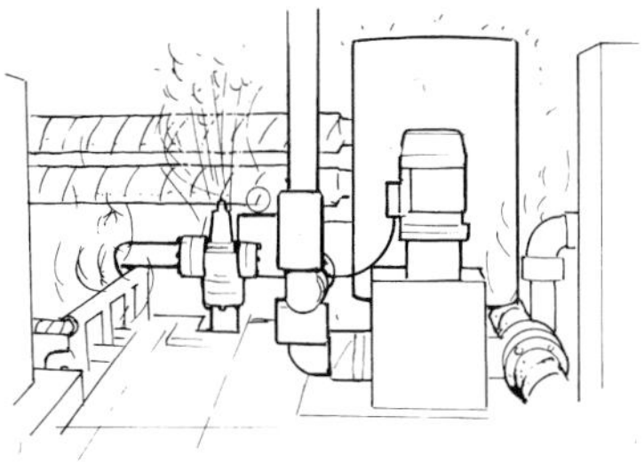

Hazardous surfaces
Hot and cold surfaces, projecting parts of equipment (particularly at head and ankle level), sharp edges and ends, tripping and slipping.

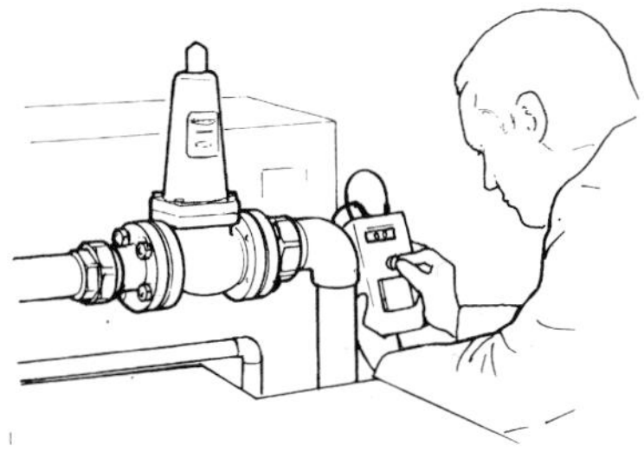

Pressurised plant and systems
Refrigeration plant, steam, compressed air systems, medical gases.
Decommissioning.

Fixings
Installation and maintenance of equipment fixing vertical, horizontal.

A GENERAL PLANNING
B EXCAVATIONS AND FOUNDATIONS
C PRIMARY STRUCTURE
D 7 BUILDING ELEMENTS AND BUILDING SERVICES
E CIVIL ENGINEERING

Group D – Building elements and building services
7 – Mechanical services

Specific hazard identification

Possible key considerations:

- **Working height** - How is the installation, commissioning and maintenance of equipment to be carried out?
- **Access for installation, commissioning and maintenance** - What is the frequency at which tasks are carried out (eg daily, monthly)?
- **Adequate space** - What are the provisions for mechanical or manual handling of equipment in the plant room and along ingress/egress routes?
- **Hazardous substances** - How will use and storage take place in the installation and operation of mechanical services (eg refrigerants ammonia, propane, water treatment chemicals, fine fibres from insulants, microbiological contamination of water services, etc.)?
- **Maintenance** - What are the provisions for the removal and replacement of plant items.
- **Electrical supply to mechanical services**
- **Any others?**

Prompts

- Rooftop locations
- Tripping hazard - low level pipework in plant rooms
- Hot surfaces - boilers and pipework
- Cold surfaces - liquefied gas pipework
- Sharp surfaces - air-handling plant casing, ductwork cladding
- Projecting parts - valve stems and wheels, head level pipes and ducts, etc.
- Pressure systems
- Manual mechanical handling
- Rotating machinery
- Safety valves
- Fixing details
- Welding vs mechanical fixings

Hazards consideration in design

Stage	Considerations and issues	Possible design options to avoid or mitigate hazards identified for mechanical services
Concept design	• Heating/cooling/ventilation strategy • Plant room sizes and locations • Principal distribution routes • Fuel sources and storage • Existing site services	• A building with good thermal characteristics (eg external envelope) will often reduce the size and complexity of the mechanical services required. • The handling of large plant items and the amount of space can be reduced by considering modular or decentralised plant. • Location of major plant and main distribution routes for pipes and ducts need careful planning and coordination, particularly riser cores to avoid congestion.
Scheme design	• Installation strategy • Plant duties and nominal sizes and weights • Pipe and duct distribution • Plant access/egress • Working space for work at all phases • Plant room layouts	• Eliminate or reduce common hazards by thoughtful selection of major plant items - falling from heights (eg low level plant), hazardous substances (eg non-toxic, non-flammable refrigerants), avoidance of pressurised systems (eg low temperature hot water heating systems). • Plan safe handling of plant components along entire access/egress routes. Locate plant rooms to keep these routes as simple as possible, (eg heavy and bulky objects at ground level - large boilers, chillers, storage vessels, etc.).
Detail design	• Handling methods for installation, commissioning and maintenance • Temporary access for installation • Materials specification • Storage of plant and materials • Safety interlocks • Asbestos • Hot working	• Layout plant rooms with an understanding of the space needed to install, commission and maintain items of plant. Provide ready access to plant items which require regular/frequent maintenance. Provide permanent guarded stairs and catwalks if this is at high level. • Consider the manual handling tasks at all phases: - can the task be carried out by one person? - will mechanical handling devices be needed? • Layout plantroom to avoid tripping hazards and projecting parts by keeping routes free of pipes and ducts. Plan distribution at high level to reduce low level pipework. Provide step-overs for low level pipes where necessary.

Examples of risk mitigation (methods of solution)

ACTION \ ISSUE	Working at height	Manual handling
Avoidance Design to avoid identified hazards but beware of introducing others	Design to install and maintain from floor level.	Use modular plant which can be easily dismantled.
Reduction Design to reduce identified hazards but beware of increasing others	Key components to be accessed from floor level.	Provide sufficient access and space around the plant.
Control Design to provide acceptable safeguards for all remaining identified hazards	Provide protected catwalks stairs and ladders.	Provide lifting beams for mechanical lifting.

Examples of risk mitigation (issues addressed at different stages)

Installation of chiller plant to serve an air conditioned department store with substantial heat gain and long run hours.

Concept design	Scheme design	Detail design
Consider using chillers with air-cooled condensers to avoid contamination of condenser water and handling of water treatment chemicals. Consider using HFC refridgerant to avoid using more toxic or inflammable refrigerant options.	Provide adequate space/access for withdrawal of condenser or evaporator tubes. Access/egress routes to allow the safe handling.	Detail plant room layout to reduce the amount of low-level pipe connecting to the evaporator and condenser. Specify lifting beam above chillers to facilitate removal of heavy plant components. Highlight the risk created by refrigerant leakage.

Distribution pipework and ductwork serving a four pipe fan coil system in a new 10 storey office building.

Concept design	Scheme design	Detail design
Consider access space for installation and maintenance at the cores where several services are located, eg ductwork, pipework, electrics, lifts, etc. Segregate service risers to avoid electric shock, fire, explosion hazards.	Co-ordination of horizontal pipework and ductwork in false ceiling voids with structure and other services to provide sufficient space for fixing of brackets for installation and maintenance. Provide clear access to regularly maintained items such as fan coil units, fitters, fire dampers, etc.	Use screwed or flanged connections on pipework to avoid welding. Consider as much prefabrication of pipework and ductwork as possible to reduce the amount work at height.

Related issues

References within this document

References	Related issues
A1	Surrounding environment determine some effects on and from this work.
A4	Access for working at height in an area of existing plant needs clear access and egress routes/systems.
D2	Roofs are often used to house mechanical plant.
D6	Cleaning of buildings will include and/or use mechanical services.
D8	Electrical services almost always interface with mechanical services.
D10	Lifts, escalators and auto walks may also be associated with mechanical services.
E8	Pipes and cables (forming the external infrastructure) will interface with mechanical services.

References for further guidance

Primary general references and background information given in Section 4

Primary

- Ministry of Defence, Defence Works Functional Standard, Design and Maintenance Guide 08. Space requirements for plant access, operation and maintenance 1996
- BSRIA Technical Note TN 9/92 – Space and weight allowances for building services plant: Inception stage design 1993
- BSRIA Technical Note TN10/92 – Space allowance for building services: Detail design stage 1992
- BSRIA Application Guide 11/92 – Design for maintainability 1992
- Department of Health NHS Estates, Health Technical Memorandum (HTM) 2023 – Access and accommodation for engineering services 1995
- Department of Health, NHS Estates, Health Technical Memorandum (HTM) 2040 – The control of legionellae in healthcare premises 1993

Secondary

- HSE PM5 Automatically controlled steam and hot water boilers
- HSE L8 The control of legionella bacteria in water systems. Approved code of practice and guidance

Background

- BS 8313:1997 Code of practice for accommodation of building services in ducts

Classification

CAWS Group S&T, **CEWS** Class Z, **CI/SfB** Code (54)/(59)

Group D – Building elements and building services
8 – Electrical services

Read the Introduction before using the following guidelines

Scope

- HV equipment and transformers.
- LV switchgear and components.
- Uninterruptible power supplies (UPS).
- Generators.
- Lightning protection.
- Fire and security alarm systems.
- Lighting and power cabling.
- Emergency lighting.
- Mechanical plant supplies.

Exclusions

- Large-scale power generation plant and distribution systems.
- Areas designated as having potentially explosive atmospheres - "hazardous areas".
- Temporary services during construction.

Major hazards

Refer to the Introduction for details of accident types and health risks

Confined working spaces
Installation, maintenance, commissioning.

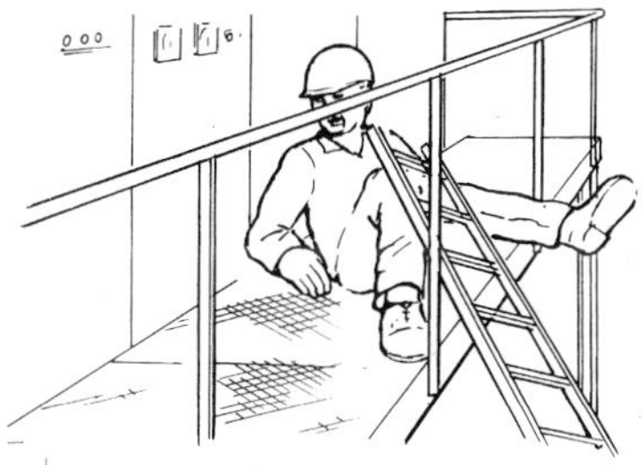

Falling from heights
Installation and maintenance of equipment, falling into pits (eg switchrooms). Builders work openings.

Manual handling of plant
Large and heavy switchgear components, transformers, generators, etc.

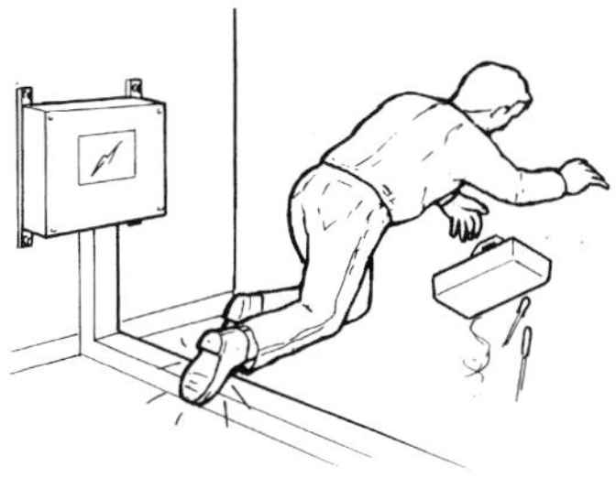

Hazardous surfaces
Hot surfaces, projecting parts of equipment (particularly at head and ankle level), sharp edges and ends, tripping and slipping.

Hazardous atmospheres/ substances
Insulating oils from transformers or HV switchgear, acids from batteries, hydrogen from charging batteries, etc.

Fixings
Installation and maintenance of fixings for equipment, on walls or in floor/ceiling zones.

Group D – Building elements and building services
8 – Electrical services

Specific hazard identification

Possible key considerations:

- **New build or refurbishment** – identify if the power supply to the building will be "live" during any electrical works?
- **Working height** – how is the installation and maintenance of equipment to be carried out, (eg replacement of lamps)?
- **Access for installation and maintenance** - What is the frequency needed for access to equipment (daily, monthly, etc.)?
- **Adequate space** – what are the provisions for mechanical/ manual handling of equipment in equipment rooms and along access/egress routes through the building?
- **Environment** – could there be flammable or explosive gases or dusts in the space?
- **Maintenance** – what are the provisions for removal and replacement of equipment.
- **Any others ?**

Prompts

- Equipment located at high level or in false ceiling voids
- Sharp surfaces – electrical tray work and conduits.
- Manual handling (eg removal of switchgear truck)
- Hazardous substances (eg transformer oil)
- Hot surfaces – generator engines and flues
- Enclosure/protection
- Fixing details

Hazards consideration in design

Stage	Considerations and issues	Possible design options to avoid or mitigate hazards identified for electrical services
Concept design	• Electrical supply strategy • Equipment room sizes and locations • Construction requirements - small plant, lighting and plant, UPS • Principal distribution routes • Existing site services • Installation strategy • Equipment and switchgear capacity, nominal sizes and weights	• Maximum size of low-voltage supply may be determined by electricity company. Maintenance of high voltage systems require specially-certificated personnel. • Locate main equipment rooms at outer walls of buildings, where possible, to allow heavy equipment to be placed in one lift from lorry to final position. • Locate equipment away from explosive/flammable atmospheres • Locate principal distribution routes away from plant rooms used by other services (if possible). • Route external cable trenches away from existing routes to avoid coincident excavations. • Avoid whenever possible carrying out electrical work on "live" services.
Scheme design	• Main risers and sub-circuit routes • Plant access/egress routes • Switchgear and equipment room layouts	• Ensure sufficient space to install largest item of plant safely. • Use dedicated risers and routes for electrical services wherever possible so the electrical installers are not put in danger from other (unfamiliar) services.
Detail design	• Handling methods for installation and maintenance • Temporary access for installation • Materials specification • Separation of services • Earthing and bonding • Enclosure/protection	• Locate equipment with sufficient working space so emergency egress is not obstructed. • Provide lifting beams where heavy parts need to be lifted for maintenance. • Ensure any temporary access for installation does not require electrical contractor to dismantle pipes or rebuild a wall. • Wherever possible, avoid materials that may require special handling or protection during disposal. • Specify all equipment to avoid risk of fire/explosion

Examples of risk mitigation (methods of solution)

ACTION / ISSUE	Hazardous substances	Manual handling during installation
Avoidance Design to avoid identified hazards but beware of introducing others	Consider use of non combustible materials for all electrical equipment to guard against electrical/construction fires.	Design modular switchboards with smaller components.
Reduction Design to reduce identified hazards but beware of increasing others	Specify fire retardant/low smoke insulants on cables instead of PVC.	Design access routes with sufficient space for mechanical handling devices.
Control Design to provide acceptable safeguards for all remaining identified hazards	Co-ordinate with other designers to incorporate smoke vents/ventilation in areas where work will take place.	Incorporate lifting beams for electrical gear into primary structure.

Examples of risk mitigation (issues addressed at different stages)

In designing the mains supply to an urban office building it becomes apparent that building loading requires a new HV supply. The only space available for transformer is in the basement adjacent to the car park area.

Concept design	Scheme design	Detail design
Location presents considerable fire risk so use a dry transformer with SF6 or vacuum switchgear. Put all cabling at high level. Access to the new switchroom will be via the car park - incorporate security measures to prevent unauthorised access.	Design to allow HV switchgear and transformer to be installed together in one room. Allow space for removal of HV switchgear and rear access to LV switchgear.	Specify sealed batteries for HV backup to avoid hazard of explosive gasses from charging unsealed batteries.

Large incoming supply arrangement with multiple transformers and both high voltage and low voltage switchgear.

Concept design	Scheme design	Detail design
Consider installing transformers in separate vaults to allow safe working on one while others are still energised. Provide separate rooms for LV and HV switchgear.	Design system with comprehensive electrical interlocking and earthing facilities to ensure no parts can become live unexpectedly. Ensure the interlocks prevent the possibility of earthing live sections or "backfeeding" through the transformers.	Ensure specification of all switchgear includes security provisions, eg padlocks.

A GENERAL PLANNING
B EXCAVATIONS AND FOUNDATIONS
C PRIMARY STRUCTURE
D 8 BUILDING ELEMENTS AND BUILDING SERVICES
E CIVIL ENGINEERING

Related issues

References within this document

References	Related issues
A4	Access is a crucial issue for installing or maintaining electrical services.
D7	Mechanical services will (almost) always require electrical services.
D10	Lifts and escalators will be electrically driven.
E8	Pipes and cables from external infrastructure will interface with electrical services.

References for further guidance

Primary general references and background information given in Section 4

Primary

- Ministry of Defence, Defence Works Functional Standard, Design and Maintenance Guide 08. Space requirements for plant access, operation and maintenance 1996
- BSRIA Technical Note TN 9/92 – Space and weight allowances for building services plant: Inception stage design 1993
- BSRIA Technical Note TN10/92 – Space allowance for building services: Detail design stage 1992 BSRIA Application Guide 11/92 – Design for maintainability 1992
- Department of Health NHS Estates, Health Technical Memorandum (HTM) 2023 – Access and accommodation for engineering services 1995
- Department of Health, NHS Estates, Health Technical Memorandum (HTM) 2040 – The control of legionellae in healthcare premises 1993

Secondary

- HSE HSR25 Memorandum of guidance on the Electricity at Work Regulations
- BS 8313:1989 Code of practice for accommodation of building services in ducts

Background

- BS 7671:1992 Requirements for Electrical Installations
- Stokes, G (1999) A practical guide to the wiring regulations. 2nd ed (Blackwell Science)

Classification

CAWS Group V, **CEWS** Class Z, **CI/SfB** Code (6-)

Group D – Building elements and building services
9 – Public health services

D9

Read the Introduction before using the following guidelines

Scope

- Cold water supply.
- Hot water supply.
- Water treatment plant.
- Surface water and foul drainage.
- Waste and refuse disposal systems.
- Above - ground sanitary systems, soil waste and vents.
- Water and gas fire extinguishing systems.

Exclusions

- Commercial incinerator plant.
- Chemical waste and drainage systems.
- Sewage treatment plants.

Major hazards

Refer to the Introduction for details of accident types and health risks

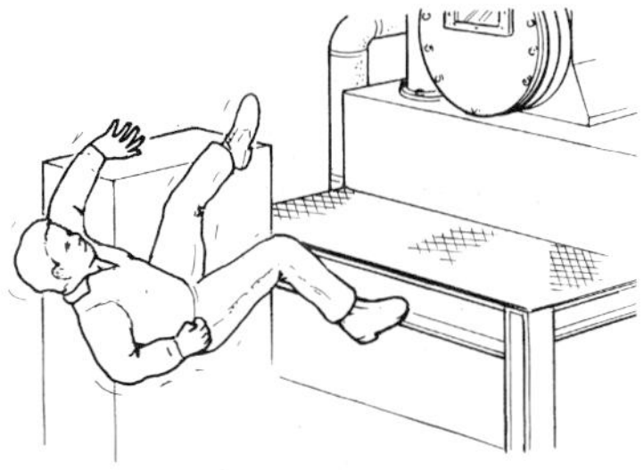

Falls from heights
During installation, commissioning and maintenance of equipment. Falling into excavations.

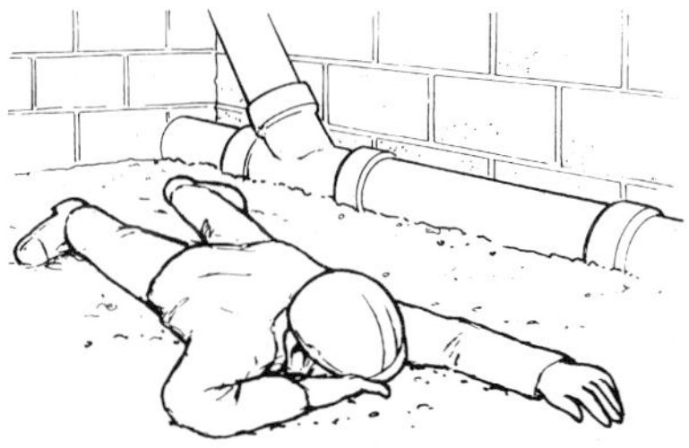

Deep excavations
Collapse of excavation, working in confined spaces, hazardous gases.

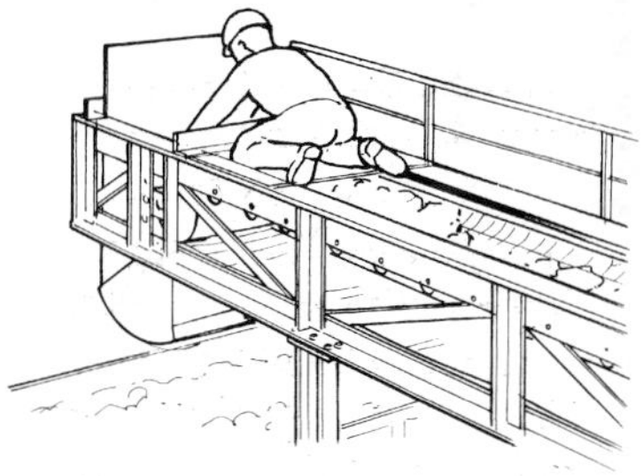

Access
Routes to equipmnt.
High level equipment.
Equipment in holes or tunnels.

Manual handling
Tanks and other public health plant are generally large and heavy or bulky.

Hazardous atmospheres/substances
Explosive atmospheres
(e.g methane), gas, legionella.
Suffocating atmospheres
(eg carbon dioxide).
Poisonous atmospheres
(eg hydrogen sulphide).

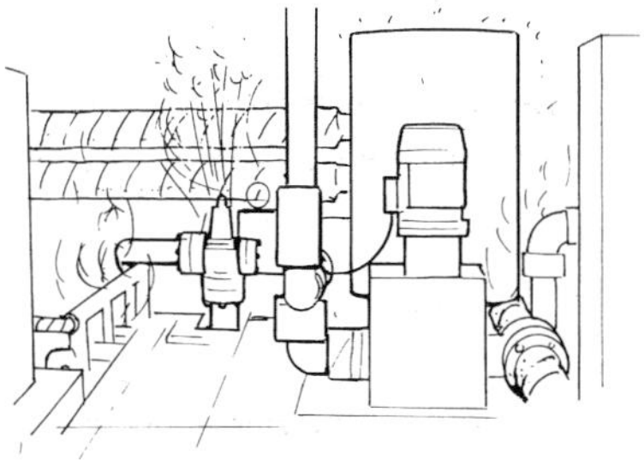

Hazardous surfaces
Projecting parts of equipment (particularly at head and ankle level), sharp edges and ends, tripping and slipping.

Specific hazard identification

Possible key considerations:

- **Working height** - How is the installation and maintenance of equipment to be carried out?
- **Access for installation and maintenance** - What is the frequency at which tasks are carried out, (eg daily, monthly, etc.)?
- **Adequate space** - What are the provisions for mechanical or manual handling of equipment in the plant room and along access/egress routes through the building?
- **Hazardous substances** - How will the use and storage take place in the installation and operation of mechanical services (eg water treatment chemicals, fine fibres from insulants, etc.)? Is there risk of microbiological contamination of water services?
- **Maintenance** - What are the provisions for the removal and replacement of items of plant?
- **Any others?**

Prompts

- Roof top locations
- Tripping hazard - low - level pipework in plant rooms
- Hot surfaces - calorifiers and pipework
- Sharp surfaces - cladding
- Projecting parts - valve stems and wheels, head level pipes and ducts, etc.
- Pressure systems
- Manual handling.
- Fixing details
- Purge existing systems

Hazards consideration in design

Stage	Considerations and issues	Possible design options to avoid or mitigate hazards identified for public health services
Concept design	• Water supply/waste/fire extinguishing - systems strategies • Plant room sizes and locations • Principal distribution routes of all services • Water sources and storage • Existing site services • Installation strategy • Plant weights • Plant access/egress routes • Working space for installation • Handling methods for installation and maintenance • Materials specification	• Locate bulk water storage tanks in plant rooms with minimum heat gain. Establish principal distribution routes which allow safe manoeuvring of large diameter heavy pipework, during installation and renewal. • Allow access around proposed plant when assessing plant room sizes. These can be refined later. • Establish preliminary information on existing site services.
Scheme design		• Specify sectional tanks for bulk water storage, for ease of handling. • Allow working space around plant and tanks for installation and maintenance. • Provide working platforms in risers where floor slab is not continued through. • Locate all existing site services before laying out external mains and drainage. Keep excavations clear of existing structures and services. • Minimise depth of excavations
Detail design		• Consider access for lifting equipment which may be required for heavy high level services. • Specify drainage sump pumps which can be maintained at floor level. • Consider (lightweight) plastic materials for larger diameter high-level pipework. • Locate drainage access points where they can be easily and safely maintained. • Locate valves on pipelines at points of safe permanent access. • Ensure safe routes available for replacement of plant. • Holes in GRP tanks to be formed at works to avoid drilling on site.

Examples of risk mitigation (methods of solution)

ACTION \ ISSUE	Installing plant at high level	Manual handling during dismantling
Avoidance Design to avoid identified hazards but beware of introducing others	Design to install and maintain plant at floor level.	Use modular plant which can be dismantled into manageable sections.
Reduction Design to reduce identified hazards but beware of increasing others	Key components to be accessed from floor level.	Provide sufficient access and space around the plant to allow mechanical handling devices to be used.
Control Design to provide acceptable safeguards for all remaining identified hazards	Provide protected catwalks stairs and ladders to high level plant.	Provide lifting eyes and/or runways for lifting plant with manual assistance.

Examples of risk mitigation (issues addressed at different stages)

Large diameter drainage run is to be routed through 5m high plantroom at high level.

Concept design	Scheme design	Detail design
Plan route which will have access from below for lifting equipment. This could involve mechanical lifting devices. Avoid routes over arears where future plant might be installed to aid replacement	Consider plastic pipes rather than cast iron, to ease manual handling. For cast iron pipes, specify mechanical joints.	Locate cleaning access doors where they can be accessed from a gantry or permanent catwalk. Where possible, avoid high level cleaning access points by extending access through the floor slab and carry out maintenance from the floor above.

Drainage outfall from a building is along a main traffic route from office car park.

Concept design	Scheme design	Detail design
Route drainage (where possible) below pavement or landscaping. Avoid deep excavation adjacent to existing structures.	Locate manholes away from site roads. Where manholes cannot be avoided in roadway, locate them to one side. If sufficient depth is available, use "side entry" manhole with access shaft in pavement.	In deep manholes use ladders rather than step irons for safer access. Where large heavy duty covers are required, consider multiple covers to reduce manual handling load. Where outfall pipe from manhole is 600mm diameter or greater, fit safety chain across outlet.

Group D – Building elements and building services
9 – Public health services

Related issues

References within this document

References	Related issues
A4	Access for maintenance and commissioning.
B2	Deep basements and shafts.
B3	Trenches for services.
D2	Roofs - rainwater drainage.
D8	Electrical services connect to equipment and plant.
E8	Pipes and cables.

References for further guidance

Primary general references and background information given in Section 4

Primary

- Ministry of Defence, Defence Works Functional Standard, Design and Maintenance Guide 08. Space requirements for plant access, operation and maintenance 1996
- BSRIA Technical Note TN 9/92 – Space and weight allowances for building services plant: Inception stage design 1993
- BSRIA Technical Note TN10/92 – Space allowance for building services: Detail design stage 1992 BSRIA Application Guide 11/92 – Design for maintainability 1992
- Department of Health NHS Estates, Health Technical Memorandum (HTM) 2023 – Access and accommodation for engineering services 1995
- Department of Health, NHS Estates, Health Technical Memorandum (HTM) 2040 – The control of legionellae in healthcare premises 1993
- HSE HSG70 Control of legionellosis

Secondary

- BS 8313:1989 Code of practice for accommodation of building services in ducts

Background

- BS 8005:1987 Sewerage Part 1 Guide to new sewerage construction
- BS 8301: 985 Code of practice for building drainage
- Building Regulations 1991. Approved document G – Sanitary conveniences and washing facilities, bathrooms and HWS storage 1992
- BS 6700:1997 Services supplying water for domestic use within buildings
- Institute of Plumbing (2002) Plumbing engineering services design guide (Institute of Plumbing)
- WRc (2001) Sewers for adoption (WRc)

Classification

CAWS Group R&S, **CEWS** Class Z, **CI/SfB** Codes (52)/(53)

Group D – Building elements and building services
10 – Lifts, escalators and autowalks

D10

Read the Introduction before using the following guidelines

Scope

- Traction lifts.
- Hydraulic lifts.
- Escalators.
- Autowalks.

Exclusions

- Temporary mobile platforms, scissor lifts.
- General mechanical handling equipment, eg document handling, baggage handling.

Major hazards

Refer to the Introduction for details of accident types and health risks

Collapse
Heavy equipment on un-propped floors.

Falling from heights
Falling into pits and shafts. Access to lift motor room.
Installation, commissioning and maintenance of equipment.

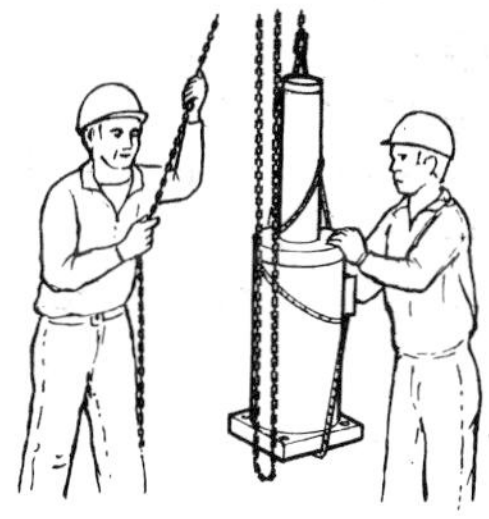

Manual handling
Large and heavy items of equipment.
Space for installation, commissioning and maintenance.

Hazardous surfaces
Projecting parts of equipment (particularly at head and ankle level), sharp edges and ends, tripping and slipping.

Hazardous substances
Dusts and vapours, bottled gases, fibrous dust, asbestos, hydraulic oils, etc.
Working in confined spaces.

Installation
Noise and vibration, entanglement, trapping, crushing, cutting by moving parts.

Specific hazard identification

Possible key considerations:

- **Insulation, isolation, overload protection and earthing of equipment** - What provisions have been made?
- **Working height** - How is the installation and maintenance of equipment to be carried out?
- **Access for installation, commissioning and maintenance** - What is the frequency at which access will be required to equipment (daily, monthly)?
- **Adequate space** - What are provisions for mechanical or manual handling of equipment in equipment rooms and along ingress/egress routes through the building?
- **Maintenance** - What are the provisions for the removal and replacement of items of equipment?
- **Hazardous substances** - What provision has been made for the use, movement and storage of hazardous substances?
- **Any others?**

Prompts

- Working at high level
- Working in lift shafts
- Sharp surfaces - equipment casing, cladding, etc.
- Manual handling
- Hazardous substances eg hydraulic oil
- Projecting parts - valves, wheels, pipes and ducts
- Fire risks - shafts and machine rooms
- Hot surfaces - motors and hydraulic tanks

Hazards consideration in design

Stage	Considerations and issues	Possible design options to avoid or mitigate hazards identified for lifts, escalators, etc.
Concept design	• Location of lifts and motor rooms • Scope of works • Familiar work • Access • Installation sequences • Lift equipment types • Handling of equipment components • Protection • Installation, commissioning and maintenance procedures • Space for services • Handling of equipment • Finishes • Positioning • Temporary supports	• Consider the time-scale of operation and delivery of plant, avoid duplication of tasks such as striking and re-erection of scaffolding in lift wells. • Consider alternative routes for access. Provide sufficient working space for installation, commissioning and maintenance.
Scheme design		• Specify equipment that is compact in design to allow for easy handling and placing in position. • Allow for suitable permanent/temporary lifting facilities (marked with SWL) for final positioning of plant. • Consider components which are pre-fabricated and delivered to site as packaged units. • Include provision for delivery and handling of components on congested sites.
Detail design		• Allow for future repair or renewal of equipment, by providing suitable lifting access/egress routes for plant and personnel. • Allow for adequate safety harness attachments for personnel working at heights within the lift well. • If mechanical handling is necessary, incorporate into the design provision for mechanical lifting and handling.

Examples of risk mitigation (methods of solution)

ACTION	ISSUE: Maintenance in confined spaces
Avoidance Design to avoid identified hazards but beware of introducing others	Specify long life, sealed components: These to be replaced rather than maintained.
Reduction Design to reduce identified hazards but beware of increasing others	Specify minimal maintenance components.
Control Design to provide acceptable safeguards for all remaining identified hazards	Provide sufficient space for maintenance tasks.

Examples of risk mitigation (issues addressed at different stages)

The installation of an escalator into a building. A compact escalator system is proposed which can be delivered in one piece to reduce programme requirements.

Concept design	Scheme design	Detail design
The overall concept for the method of installation should be planned.	Ensure adequate space allowed for installation.	Further consideration of the sequence of installation. Identify specific hazards due to the building structure.

Panoramic lift installation to serve a department store

Concept design	Scheme design	Detail design
The lift should be designed so that it can be installed in a completed form, or consideration to be given to access/egress routes to allow safe handling of components.	Panoramic lifts use laminated glass, for walls, ceilings and doors. Provision should be made for the safe handling of this material and future cleaning requirements.	The design should allow for safe cleaning of the glass. Allocate one floor, (eg ground level), where all-round access can be arranged. Ensure adequate depth of well to allow safe maintenance.

A GENERAL PLANNING
B EXCAVATIONS AND FOUNDATIONS
C PRIMARY STRUCTURE
D 10 BUILDING ELEMENTS AND BUILDING SERVICES
E CIVIL ENGINEERING

Related issues

References within this document

References	Related issues
D3	Roofs may house lift motor rooms.
D7	Mechanical services interact closely with the installation and maintenance of vertical transportation components.
D8	Electrical services need to be considered for the installation and maintenance of lifts and escalators, also emergency lighting.
D9	Public health services - often impinge on the installation and maintenance of lifts and escalators.
	Fire engineering needs to be considered for fire-fighting lifts.

References for further guidance

Primary general references and background information given in Section 4

Primary

- BS EN115:1995 Safety rules for the construction and installation of escalators and passenger conveyors
- National Association of Lift Manufacturers. Principles of planning and programming a lift installation
- National Association of Lift Manufacturers/CIBSE – Guidance on the management and maintenance of lifts and escalators

Secondary

- CIBSE Guide D Transportation in buildings 1993.

Background

- BS 5655 Lifts and service lifts: Parts 3 to 12 inc Part 6 (2002) Selection and installation of new lifts
- BS 7255: 2001 Code of practice for the safe working on lifts
- BS 5588: Part 5: 1991 Code of practice for fire fighting stairs and lifts
- Strakosch, G R (1998) The vertical transportation handbook (John Wiley)

Classification

CAWS Group X, **CI//SfB** Code (66)

Group E – Civil engineering
1 – General civil engineering – including small works

Read the Introduction before using the following guidelines

Scope

- Civil engineering construction sites.
- More detail on selected civil engineering operations is given in work sectors E2 to E9.

Exclusions

- Demolition (see A2)

Major hazards

Refer to the Introduction for details of accident types and health risks

Buried/crushed/trapped
Collapsed excavation of service trench/foundation.
Backfilling in confined space.
Falsework/formwork collapse.

Falls
Awkward access/egress.
Inadequate working space.
Unstable platforms.

Health hazards
Chemicals.
Asbestos.
Gases.
Water-borne diseases and contamination.
Contaminated ground.
Manual handling.

Machinery and transport
Site traffic and plant.
Public highway traffic adjacent to site.
Unloading materials and equipment.

Services
Power cables (underground and overhead).
Fuel and gas pipelines. Sewers.
High pressure mains water.

Ground movement
Landslips, flooding.
Subsidence.
Fractured services.

E1 Group E – Civil engineering
1 – General civil engineering – including small works

Specific hazard identification

Possible key considerations:

- **Site survey** - What is the history of site? Is there contamination from previous use?
- **Ground survey** - What significant foundation problems will be encountered?
- **Working room** - What is the most appropriate plant?
- **Flooding** - What is the likelihood of a flood occurring?
- **Services** - Where are they? What measures to divert or protect?
- **Any others?**

Prompts

- Previous site use
- Waste disposal
- Access/egress
- Adjacent property
- Adjacent operations
- Construction planning
- Noise/vibration/dust.
- On-site storag
- Confined spaces/ventilation

Hazards consideration in design

Stage	Considerations and issues	Possible design options to avoid or mitigate hazards identified for general civil engineering
Concept design	• Why this site? • Site history • Affected local buildings • Depth of foundations • Location of services • Access routes • Ground conditions • Topography • Drainage/hydrology • Working space • Plant/machinery • Excavation support • Spoil/contamination • Drainage • Advance works • Protection of public • Maintenance • Cleaning	• Review the practicalities of other layouts or routes with fewer/less serious hazards. • Choose a massing arrangement of structures which will allow wide battered excavations. • Design to consider effect on local environment (eg access/egress, noise and vibration, lighting), not just the convenience and value of the project. • Assess the possible benefits from further site investigation or specialist surveys. • Avoid disturbance of existing services.
Scheme design		• Consider drainage measures to improve the stability of excavations. • Ensure that construction methods are practical in relation to the working space available.
Detail design		• Design to minimise excavation depths. • Consider appropriate methods for groundwater control. • Consider an enabling works contract to protect or divert services. • Include "systems" for future safe working procedures for maintenance, cleaning and possibly demolition phases.

Examples of risk mitigation (methods of solution)

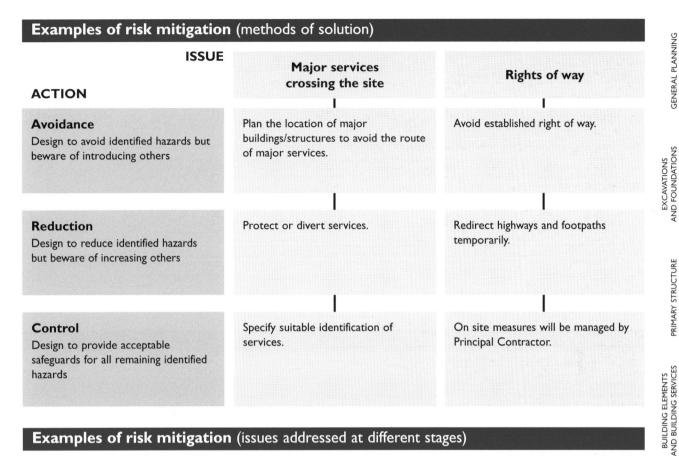

ACTION \ ISSUE	Major services crossing the site	Rights of way
Avoidance Design to avoid identified hazards but beware of introducing others	Plan the location of major buildings/structures to avoid the route of major services.	Avoid established right of way.
Reduction Design to reduce identified hazards but beware of increasing others	Protect or divert services.	Redirect highways and footpaths temporarily.
Control Design to provide acceptable safeguards for all remaining identified hazards	Specify suitable identification of services.	On site measures will be managed by Principal Contractor.

Examples of risk mitigation (issues addressed at different stages)

A small group of residential dwellings is to be constructed on the site of a disused small sewage works, adjacent to a river. Proposals are required to deal with the residual site contaminants.

Concept design	Scheme design	Detail design
Complete site inspection to identify contamination in relation to new dwellings. Identify history and construction records of site. Locate services.	Confirm extent of contamination and ground water details by site investigation. Consider methods to contain/remove contaminants. Identify ground dewatering methods, excavation, drainage requirements, etc.	Develop details for foundations and drainage which minimise ground disturbance.

A flood alleviation scheme requires surface water storage capacity under a residential street.

Concept design	Scheme design	Detail design
Find the optimum balance between cross-sectional area and length of sewer to allow construction access while retaining pedestrian access to the houses.	Determine the position in the road which minimises the disruption and interfaces with existing services.	Locate penstock taps and main chamber access covers in a traffic island or wide pavement for safer access during maintenance.

Group E – Civil engineering
1 – General civil engineering – including small works

A GENERAL PLANNING
B EXCAVATIONS AND FOUNDATIONS
C PRIMARY STRUCTURE
D BUILDING ELEMENTS AND BUILDING SERVICES
E1 CIVIL ENGINEERING

Related issues

References within this document

References	Related issues
A1	Local environment protection is required from the hazards associated with any construction work.
A2	Site clearance and demolition
A3	Ground investigations should be related to the project to be constructed.
A4	Access/egress routes may affect the site and hence the construction materials to be used and the frequency of deliveries.
A5	Site layout will be determined by the project to be constructed on the site available and plant to be used. If site restricted, hazards of construction could be accentuated.
C1-C4	Concrete construction
C5,C6	Structural Steelwork construction
E2,E3	Railways/roads/water, adjacent construction work should include safe distances and suitable protection for persons on site.

References for further guidance

Primary general references and background information given in Section 4

Primary

- CIRIA SP57 (1988) Handling of materials on site
- CIRIA R97 (1992) Trenching practice
- HSE HSG47 Avoiding danger from underground services
- HSE GS6 Avoidance of danger from overhead electric power lines
- HSE HSG66 Protection of workers and the general public during the development of contaminated land
- HSE CIS36 Silica

Secondary

- HSE INDG258 Safe work in confined spaces
- HSE GS289 Parts 1 to 4 health and safety in demolition work
- HSE GS28 Parts 1 to 4 Safe erection of structures
- CIRIA R130 (1993) Methane: its occurrence and hazards in construction
- CIRIA C515 (2000) Groundwater control – design and practice
- CIRIA R182 (1998) Pumping stations – design for improved buildability and maintenance
- CIRIA R185 (1999) The observational method in ground engineering: principles and applications
- CIRIA SP32 (1984) Construction over abandoned mineworkings
- HSE MDHS100 Surveying sampling and assessment of asbestos coating materials

Background

- BS 6031:1981 Code of practice for earthworks
- BS 8002:1994 Earth retaining structures
- Tomlinson, M J, Foundation design and construction (Prentice Hall)

Classification

CEWS Classes A-Z, **CI/SfB** Code (--)1

Group E – Civil engineering
2 – Roads, working adjacent to, maintenance of

Read the Introduction before using the following guidelines

Scope

- Existing roads - motorways, dual carriageways, all-purpose roads in rural, urban and city centre locations.
- New road construction, including motorway widening.
- More detail on selected civil engineering operations is given in E1 and E3 to E8.

Exclusions

- Roads in tunnels.
- Roads on viaducts.
- Temporary roads.

Major hazards

Refer to the Introduction for details of accident types and health risks

Traffic
Working adjacent to live traffic. Traffic management in temporary conditions.
Intrusion of runaway vehicles.

Falls
Equipment dropped from overhead structure/top of retaining wall.
Falls into route of traffic.

Plant and machinery
Collisions between construction plant, highway traffic and pedestrians.
Crossing and slewing over the highway.

Services
Excavation of pipes, drains and cables.
Collision with overhead cables.

Buried/crushed/trapped
Collapse of excavation for drainage or services.
Landslip triggered by excavation near toe of batter/surcharge.

Health hazards
Damaged hearing.
Vibration white finger.
Inhalation of dust and solvents.
Contaminated land.

Specific hazard identification

Possible key considerations:

- **Site survey** - What site contamination is present?
- **Ground investigation** - What water table and slope stability problems exist?
- **Access** - How is safe access and egress provided?
- **Working room** - How can sufficient space for safety zones and plant/equipment be allocated?
- **Traffic management** - How can the design assist traffic and pedestrian management?
- **General public** - What site security arrangements are required?
- **Services - Location?** What measures are needed to protect?
- **Any others?**

Prompts

- Access Routes
- Traffic diversions
- Lane closures/signage
- Deliveries/waste disposal
- Utilities/services
- Weather
- Road/street lighting
- Protect pedestrians

Hazards consideration in design

Stage	Considerations and issues	Possible design options to avoid or mitigate hazards identified for working adjacent to and maintaining roads
Concept design	• Site history • Urban/rural location • Access routes • Services • Hydrology • Lane rental • Ground conditions • Road geometry • Drainage • Plant and machinery • Working space • Sequencing • Advance works • Traffic management • Interaction with existing traffic • Pavement • Earthworks • Noise • Maintenance	• Consider whether further soil testing/investigation will enable a better prediction of temporary slope stability. • Consider location of major overhead or underground services with respect to expected working methods. • The route location and type of road should be considered in relation to safe access and delivery of materials to the site.
Scheme design		• For easier access and movement of machinery, consider advance drainage works, watercourse diversions, etc. • Check that the chosen temporary diversion of traffic causes the least trouble elsewhere. • Design to facilitate staged construction, so that traffic is not diverted elsewhere.
Detail design		• Avoid the (temporary) diversion of services in advance of works commencing. • Traffic management measures, temporary signage and lighting to be incorporated into partially complete works. • Design to reduce frequency of maintenance, particularly for the more heavily trafficked routes. • Programme the contract periods to avoid seasonal peak traffic flows.

Examples of risk mitigation (methods of solution)

ACTION / ISSUE	Working adjacent to high speed traffic	Land slide/ embankment slip
Avoidance Design to avoid identified hazards but beware of introducing others	Divert all traffic or close one carriageway.	Minimum or zero excavation required by using piled foundations and setting the structure at a higher level.
Reduction Design to reduce identified hazards but beware of increasing others	Divert Heavy Goods Vehicles (HGV) only. Add a lane closure to create a buffer zone.	Grade the embankment or cutting to a shallower angle before excavation for other works.
Control Design to provide acceptable safeguards for all remaining identified hazards	Specify concrete barriers.	Stabilise the slope by contingency methods (such as soil nailing) and drainage, as required in response to movement monitoring.

Examples of risk mitigation (issues addressed at different stages)

A stretch of uneven road surface requires emergency repair.

Concept design	Scheme design	Detail design
The exposure of workers was assessed for both temporary repair and complete carriageway reconstruction. Possible road closure periods identified.	Investigations organised ahead of repairs to reduce the possibility of extra unforeseen works. Roadwork contracts on neighbouring routes were accelerated or delayed to minimise overall disruption.	Materials were chosen which could be laid in narrow strips to give maximum flexibility of construction planning.

A single carriageway is to be converted to a two-lane dual carriageway.

Concept design	Scheme design	Detail design
Consider new routes on separate land, parallel land take or confining work to existing corridor.	Route details to consider safe access points for plant and operatives.	New road lighting designed to reduce the need for temporary lighting during construction.

A GENERAL PLANNING
B EXCAVATIONS AND FOUNDATIONS
C PRIMARY STRUCTURE
D BUILDING ELEMENTS AND BUILDING SERVICES
E2 CIVIL ENGINEERING

Related issues

References within this document

References	Related issues
A1	Local environment should be protected from the hazards associated with roadworks.
A3	Ground investigations should include testing for toxic materials, soil density and gases, as well as stability and groundwater details.
A4	Access/egress routes at the site should be kept clean, clear and suitable for emergency services.
A5	Site layout should be suitable for the working space available and plant anticipated to be used on the site.
C1-C4	Concrete materials brought to site in bulk for batching will require designated storage and batching areas.
C5,C6	Special Access routes for large prefabricated elements need to be planned.
E1	General civil engineering often involves working adjacent to roads
E4,E5	Bridge construction and maintenance
E6,E7	Working near water introduces the hazards of drowning and flooding.
E8	Pipes and cables are most often installed adjacent to roads.

References for further guidance

Primary general references and background information given in Section 4

Primary

- HA/CSS/HSE Guidance for safer temporary traffic management – as maintained on HA website
- HA/CSS/HSE Temporary traffic management on high speed roads – good working practice – as maintained on HA website
- HSE CIS36 – Silica
- DoT Safety at Streetworks and Roadworks: A Code of Practice, 1993

Secondary

- CIRIA R113 (1986) Control of a groundwater for temporary works

Background

- DoT Chapter 8 Traffic Signs Manual
- Wignall, A (1999) Roadwork theory and practice (Butterworth Heinemann)

Classification

CEWS Class R, **CI/SfB** Code (--)12

Group E – Civil engineering
3 – Railways, working adjacent to, maintenance of

Read the Introduction before using the following guidelines

Scope

- Working adjacent to, and over operational railways.
- Maintenance work on operational railways.
- More detail on selected civil engineering operations is given in work sectors E1, E2 and E4 to E6.

Note: Extensive and strict procedures apply when working in connection with operational railways to ensure safety of the workers and general public. The clear definition of the work area with respect to the line is of paramount importance. This will be determined by the respective operator (eg Railtrack, London Underground Ltd, Docklands Light Railway, etc).

Exclusions

- Railways in tunnels.
- Maintenance depots.
- Specialist trackwork and electrification work.
- Private railways.

Major hazards

Refer to the Introduction for details of accident types and health risks

Struck by moving trains
Organisation of personnel and material handling not in accordance with established safety procedures.

Lifting failures
Electrocution (overhead or third rail).
Damage to signals, signal controls and communications.

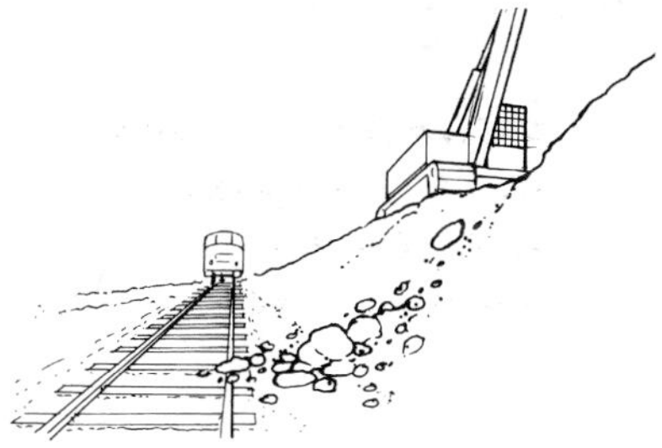

Ground instability
Rock falls into cuttings.
Embankment slips.

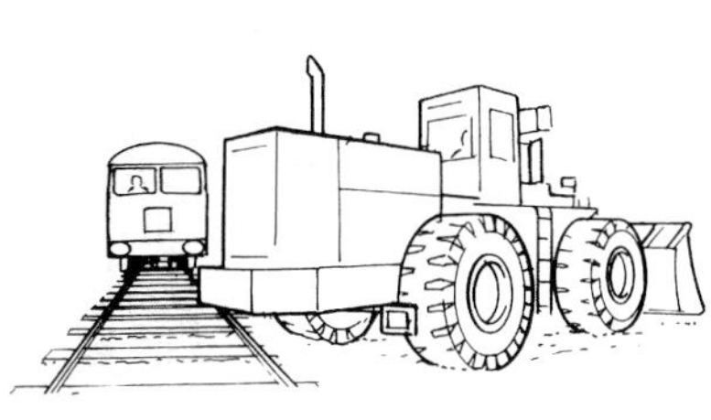

Trains collide with plant
Encroachment into permanent way envelope by plant/machinery.
Dropped equipment/components.

Falls, trips and slips
Bridge maintenance.
Uneven surfaces.
Steep cuttings and embankments. Access to services.

Health hazards
Noise and vibration.
Contaminated land.
Asbestos, oils.

Group E – Civil engineering
3 – Railways, working adjacent to, maintenance of

Specific hazard identification

Possible key considerations:

- **Electricity** - How are lines electrified (overhead and electric, 3rd or 4th Rail)?
- **Working Space** - What size of plant is to be used? Is there sufficient space?
- **Site survey/Records** - Where are cable, signals cabinets, other railway furniture located?
- **Track geometry** - How will this affect work methods?
- **Train movement** - What train movements have to be maintained?
- **Track possessions** - Will possessions be necessary or possible?
- **Track operator** - What are the requirements of the operator?
- **Operations/windows** Track possession or full closure?
- **Any others?**

Prompts

- Track working rules and procedures
- Codes of practice
- Contamination
- Protection of public
- Noise/vibration
- Weather/temperature
- Material storage
- Emergency procedures
- Access
- General public

Hazards consideration in design

Stage	Considerations and issues	Possible design options to avoid or mitigate hazards identified for working adjacent to, and working on railways
Concept design	• Historical records • Site access restrictions • Underground and overground services • Maintenance • Location • Working space • Ground conditions • Drainage • Night working • Access • Working at stations • Safe refuge areas • Disposal of waste • Cable protection • Track possessions • Delivery of materials • General Public • Other transport systems	• Consider design options in which the proximity of the railway has least effect. • The type and frequency of rail traffic may prohibit or restrict certain methods of construction. • Design for minimum track possessions or consider full closure with train diversions. • Check whether the safety of rail passengers will be impaired in anyway. Ensure separation of public and construction work.
Scheme design		• Examine the possibility of modification to cuttings and embankments to improve working space or storage areas. • Consider night working and removal of power supply to overhead lines or electrified rails. • Determine the activities which require track possession and for how long. • Provide access to control boxes etc. other than along the track. • Allow for frequent areas of safety alongside the track.
Detail design		• Consider the protection and/or duplication of signalling. equipment, controls and communications. • Match packages of work and their content to planned periods of track possession. • Highlight options for the separation of general public and work on site.

Examples of risk mitigation (methods of solution)

ACTION / ISSUE	Facility close to overhead electrification lines	Installation of inspection boxes for maintenance
Avoidance Design to avoid identified hazards but beware of introducing others	Locate facility away from overhead electric lines	Position boxes where access can be obtained away from track side
Reduction Design to reduce identified hazards but beware of increasing others	Design components to require small cranes.	Provide defined and guarded access ways to boxes alongside trackside
Control Design to provide acceptable safeguards for all remaining identified hazards	Outline erection sequences assumed. Place goal posts.	

Examples of risk mitigation (issues addressed at different stages)

A distribution warehouse for a mail-order company is to be built on a long rectangular plot of land close to an overhead electrified railway.

Concept design	Scheme design	Detail design
Design the access road to the warehouse to abut the railway land.	The designer favours more and shallower piles rather than fewer deeper piles (taller rigs). The deeper drains will be routed furthest from the railway land.	

A road crosses over a non-electrified railway line. Routine maintenance inspections of the bridge have identified that remedial works are necessary to the parapets of the bridge.

Concept design	Scheme design	Detail design
Consider track status and whether overnight working is necessary. Consult with controllers of track on requirements	Identify methods which do not encroach onto the track, even for track possession periods.	Ensure that the methods of repair have set-up, preparation, and curing periods which are compatible with the times of possession.

Related issues

References within this document

References	Related issues
A2	Site clearance activities will need to consider the local environment.
A4	Access/egress routes will also be determined by local considerations eg roads, urban/rural site interface with general public and other transport systems.
A5	Site layout will be determined by working space available and track safety measures required.
B1-B3	Excavation and earthworks will need to consider effects on drainage and permanent way track stability.
C1-C6	Construction materials may require special routes to site and careful consideration with respect to site storage.
D7-D9	Services for power conducting and signalling systems need careful attention.
E1	Working space restrictions especially on small sites accentuate problems.

References for further guidance

Primary general references and background information given in Section 4

Primary

- Railtrack GC/RT 5131 Requirements for processing outside party development proposals

Secondary

- Railtrack GO/RT3070 Track Safety Handbook 1995
- Esveld, Modern Railway Track
- Railway Group Standards

Background

- Permanent Way Institution (1995) British railway track: design, construction and maintenance (under revision)

Classification

CEWS Class A, **CI/SfB** Code (--)11

Read the Introduction before using the following guidelines

Scope

- Bridges for and over/under roads, railways, rivers, canals.
  (Includes footbridges, subways, farm bridges, temporary bridges.)
- Bridge strengthening and major repairs.

Exclusions

- Tunnels, piped watercourse, corrugated metal culverts.

Major hazards

Refer to the Introduction for details of accident types and health risks

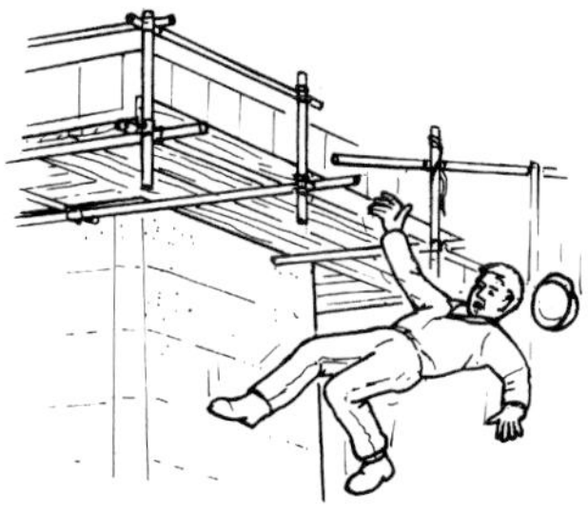

Falls
Work at height.
Inadequate access.

Intrusions
Collapse of foundation excavation/or cofferdam.
Embankment slip.
Highway traffic.

Services
Overhead and underground cables.
Pipes and cables concealed in the bridge deck.

Buried/crushed/trapped
Collapse of permanent and temporary work during construction.
Trapped under permanent bearing.

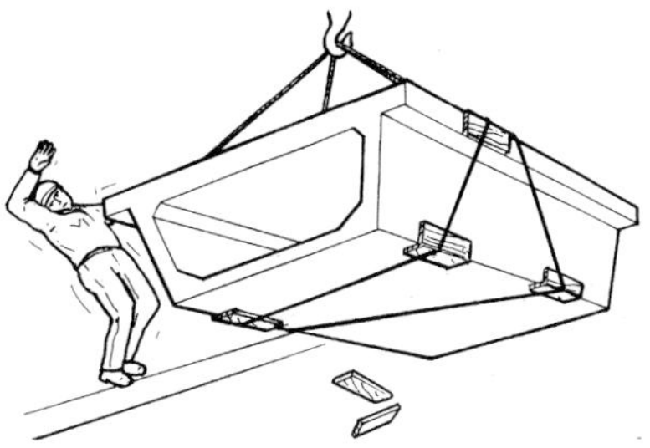

Plant and machinery
Struck by plant, or materials and units being slewed into position.
Failure of jack or cables during tensioning.

Health hazards
Concrete additives.
Liquid applied waterproofers.
Hot bitumen.
Contaminated land.

Group E – Civil engineering
4 – Bridge construction

Specific hazard identification

Possible key considerations:

- **Stability** - What procedures required for safe erection?
- **Site investigation** - What hazards do ground conditions present?
- **Access/working room** - Where can materials be stored to provide clear/safe access and working space?
- **Services** - Is the site affected by overhead and underground services? Where are the services?
- **The spanned facility** -What particular safety procedures need to be addressed for whatever is crossed?
- **Access** - What facility is required for future maintenance?
- **Public** - What protection methods are required?
- **Any others?**

Prompts

- Site history
- Access routes
- Materials delivery/storage/use
- Contaminated land
- Adjacent structures
- Demolition
- Future maintenance
- Waste disposal

Hazards consideration in design

Stage	Considerations and issues	Possible design options to avoid or mitigate hazards identified for bridge construction
Concept design	• Is a bridge necessary? • Local issues affecting design • Access to site • Overhead/underground services • Ground conditions	• Consider reducing the number of piers by increasing the length of spans. • Is the desired profile for wind resistance more safely achieved by shaping the off-site prefabricated structural units or by on-site addition of cladding? • Location of abutment to respect local ground conditions as well as alignment of routes under and over. • Choose structural form and material with respect to future
Scheme design	• Future maintenance. • Demolition • Sequence of construction • Foundations • Working space • Ground contamination • Hydrology • Services	maintenance and appropriateness of prefabrication. • Favour minimum depth of excavation for bridge foundations. • Reduce groundwater problems (that may cause instability) by appropriate drainage measures. • Avoid the need to (temporarily) divert services in advance of construction. • Consider what temporary works are expected and what construction methods are excluded by the nature of the site.
Detail design	• Materials delivery • Storage • Excavation • Temporary structural supports • Lifting measures	• More durable design solutions will reduce the frequency of future maintenance. • Ensure that prefabricated components can be handled practically by choosing symmetrical forms and incorporating lifting points. • Assess whether prefabricated units could incorporate the permanent barriers. • Include facilities for safe access to carry out future inspections and maintenance.

Examples of risk mitigation (methods of solution)

ACTION / ISSUE	Vehicle collision with abutment works	Access gantries for maintenance
Avoidance Design to avoid identified hazards but beware of introducing others	Increase span of bridge to obviate need for abutments or allow them further away from road.	Not needed because a permanent means of access is incorporated into the design.
Reduction Design to reduce identified hazards but beware of increasing others	Design a permanent barrier to be installed before abutment work commences and to remain effective throughout construction.	Access routes provided for all internal inspections and all external surfaces are reachable by lorry-mounted equipment.
Control Design to provide acceptable safeguards for all remaining identified hazards	Reduce vehicle speed and/or change routes by traffic control eg narrow lanes, closed lanes, contra-flows etc.	Brackets and holes provided at frequent intervals sufficient to construct temporary access to all areas.

Examples of risk mitigation (issues addressed at different stages)

An existing road bridge is to be reconstructed. The duration of the works is expected to be twelve months. The location is urban. Local road and pedestrian traffic at peak times is considerable. The bridge crosses a canal.

Concept design	Scheme design	Detail design
What services use the bridge as a crossing? Consider temporary closures and partial closures. Determine usage of canal and tow path.	Co-ordinate the location of a temporary bridge for service diversions, and pedestrian traffic. Bored pile design less likely to disturb canal than driven piles or large excavations.	Consider/provide suitable solution for traffic diversions and public access to properties on each side.

An existing motorway bridge is to be symmetrically widened to accommodate an additional traffic lane on either side.

Concept design	Scheme design	Detail design
Consider construction history and as-built drawings. Assess structure to determine necessary modification or demolition and replace. Additional lane and new hard shoulder? Traffic and access constraints.	Identify methods to reduce worker/road traffic conflict. Consider disruption to road traffic and identify an acceptable sequence of construction.	Identify maintenance and repair strategy for bridge. Review assumed construction methods and sequences.

Group E – Civil engineering
4 – Bridge construction

Related issues

References within this document

References	Related issues
A1	Local environment requires protection from the hazards associated with any construction work.
A2	Site clearance will involve access and demolition problems.
A3	Ground investigations should be comprehensive to allow alternative methods of construction to be investigated.
A4	Access/egress routes may affect the construction method and materials to be used at the site as well as site security and protection of the public.
A5,E1	Site layout working room will determine construction methods and appropriate construction plant to be used.
B1,B3	Excavation/foundation will need to consider drainage and stability problems.
C1-C4	Concrete construction hazards will arise in the foundations, abutments and spanning members.
C5,C6	Structural steelwork construction hazards will arise mainly for the spanning members.
C7	Masonry hazards will arise when facing abutments with stonework or similar.
C8	Timber hazards will arise for small span bridges.
E6	Working over/near water will add additional hazards for bridge construction.

References for further guidance

Primary general references and background information given in Section 4

Primary

- HSE GS28 Parts 1 to 4 Safe erection of structures
- CIRIA R155 (1996) Bridges – design for improved buildability

Secondary

- CIRIA SP57 (1988) Handling of materials on site
- Ryall, M J (2000) Bridge management 4: Inspection, maintenance, assessment and repair (Thomas Telford) and the three earlier conferences

Background

- Ryall, M J (2000) Manual of bridge engineering (Thomas Telford)
- Pritchard, B (1992) Bridge design for economy and durability (Thomas Telford)
- CIRIA SP95 (1993) The design and construction of sheet-piled cofferdams
- Highways Agency: Design Manual for roads and bridges

Classification

CI/SfB Code (--)182

Group E – Civil engineering
5 – Bridge maintenance

Read the Introduction before using the following guidelines

Scope

- Existing bridges for and over roads, railways, aqueducts, viaducts, footbridges, farm bridges, subways.

  (Maintenance to include: inspection surveys, monitoring, cleaning, and painting.)

Exclusions

- Tunnels, CM culverts, piped watercourses.
- Bridge strengthening and repairs (see E4).

Major hazards

Refer to the Introduction for details of accident types and health risks

Falls
Awkward access.
Unsafe place of work.
Difficult working position.

Plant and machinery
Use of small tools in restricted circumstances.
Struck by access gantries and lifting/slewing operations.

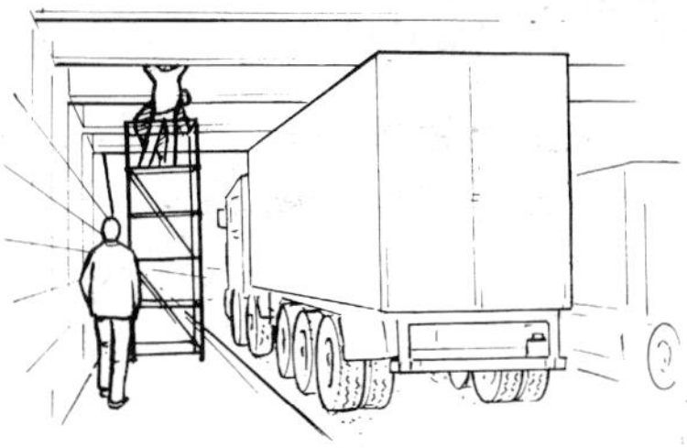

Traffic
Working adjacent road/railway traffic.
Large vehicles close to work areas under the bridge. Pedestrian routes.

Services
Cables and pipes concealed in the bridge deck.
Electrified rails and overhead cables.

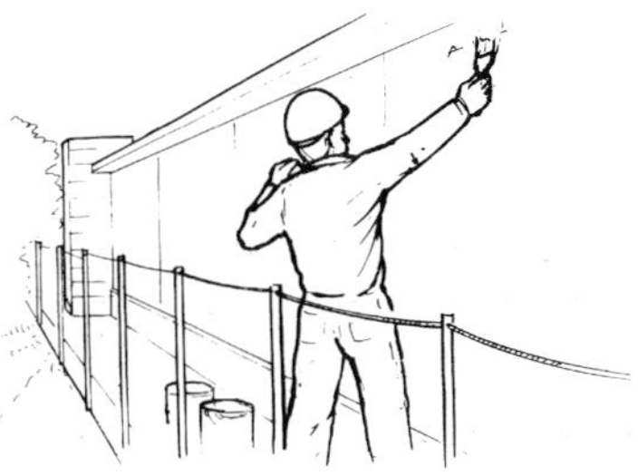

Health hazards
Waterproofing primers.
Paint and solvents. Dust. Skin irritants and allergens.

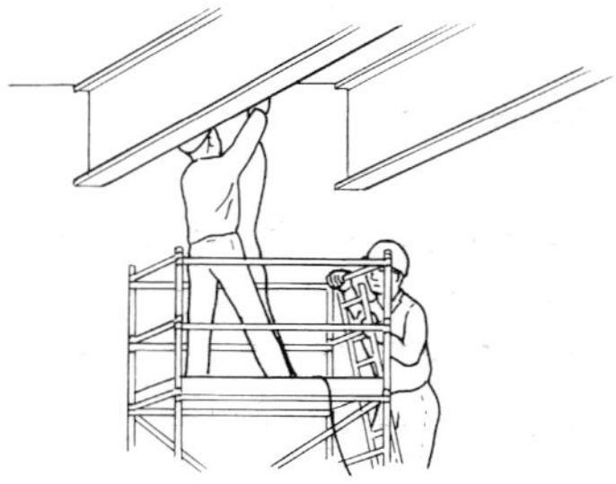

Asphyxiation
Working in confined spaces. Harmful gases.

E5 Group E – Civil engineering 5 – Bridge maintenance

Specific hazard identification

Possible key considerations:

- **Bridge location** - What is the bridge type? What does it cross (eg water, road or railway?
- **Bridge details** - Clearance, structural form, prestressed elements?
- **Access/working room** - What are suitable plant and equipment for inspection and maintenance?
- **Services** - What services exist within/without structure to be accommodated? Protection/access?
- **Environment** - Will the site or use of the bridge exclude using particular plant or materials?
- **Any others?**

Prompts

- Health and safety file?
- Construction history
- Drawings/manuals
- Ground conditions
- Working platforms
- Confined spaces
- Temporary structures
- Weather conditions
- Services
- Repair materials

Hazards consideration in design

Stage	Considerations and issues	Possible design options to avoid or mitigate hazards identified for bridge maintenance
Concept design	• Bridge type • Access • Environment • Working space • Possessions • Inspection • Ground investigation • Hard standing/platform locations • Fixtures for access assistance • Sequence/phasing • Temporary work • Hazardous materials • Plant and machinery • Services • Task lighting and visibility	• Review the implications of deferring the maintenance work. • Check demolition and replacement as an alternative to frequent maintenance. • If the extent and scope of work is uncertain, consider further inspections before advancing the design.
Scheme design		• If the work is weather sensitive, define or revise the programme accordingly. • Ensure that the preferred solution takes advantage of new developments in access equipment - eg lorry-mounted booms for access to bridge soffits.
Detail design		• Favour the selection of repair processes with less noise, dust, waste and harmful substances. • Select preparation and application methods which use remotely controlled equipment.

Examples of risk mitigation (methods of solution)

ACTION / ISSUE	Provide temporary access	Hazardous substances
Avoidance Design to avoid identified hazards but beware of introducing others	Use robots, CCTV, and electronic survey methods for remote inspection.	Exclude the use of materials, chemicals and solvents with known risks to health.
Reduction Design to reduce identified hazards but beware of increasing others	Use lorry-mounted access platforms to minimise the time period of men and equipment on site.	Hazardous substances only allowed if they significantly extend the period to first maintenance and reduce frequency of maintenance thereafter.
Control Design to provide acceptable safeguards for all remaining identified hazards	Divert traffic, calm traffic, provide barriers and buffer zones around the erection of traditional scaffolding.	Control of hazardous substances will be managed by principal contractor.

Examples of risk mitigation (issues addressed at different stages)

There is serious spalling on the soffit of a concrete bridge deck over a canal.

Concept design	Scheme design	Detail design
Determine whether the work can be effected without compromising the integrity of the bridge. Consider the practicality of using a temporary deck or pontoon for easy access.	Specify lane closures to reduce maximum load on bridge.	Select materials and methods which will allow the works to be broken down into smaller controlled areas.

Waterproofing for a road bridge deck is to minimise water ingress and hence road salt attack in order to reduce maintenance and the hazards associated with working at height.

Concept design	Scheme design	Detail design
Maximise deck cross fall with drainage facility to cater for large rainfall events. Plan bridge closure and/or traffic diversions.	Consider appropriate waterproofing options (sheet materials, liquid coatings, impregnation) and assess their inherent risks to health.	Design for eventual replacement/ ease of repair of waterproofing system.

Group E – Civil engineering
5 – Bridge maintenance

Related issues

References within this document

References	Related issues
A1	Local environment is affected by maintenance work which invariably disrupts pedestrian and road traffic movements.
A2	Site clearance relates to safe access and storage of materials.
A4	Access/egress to the structure is one of the most important safety aspects of bridge maintenance. Access constraints may also dictate plant/machinery to be used.
A5,E1	Site location/working space will dictate equipment and materials to be used. Confined working accentuates and creates hazards.
B1	Excavation will need to consider effect on drainage and bridge stability.
C1-C6	Construction materials have inherent properties and their method of application may be hazardous to health.
E6	Working over/near water will add hazards to bridge maintenance.

References for further guidance

Primary general references and background information given in Section 4

Primary

- Ryall, M J (2000) Bridge management 4: Inspection, maintenance, assessment and repair (Thomas Telford) and the three earlier conferences
- Ryall, M J (2001) Bridge management (Butterworth Heinemann)

Secondary

- HSE INDG258 Safe work in confined spaces
- CIRIA R155 (1996) Bridges – design for improved buildability

Background

- Highways Agency: Design manual for roads and bridges

Classification

CI/SfB Code (--)182 (W2)

Read the Introduction before using the following guidelines

Scope

- Inland and coastal waters.

  Including: watercourses, (rivers, canals, ditches), lakes, reservoirs, ponds, tanks, docks, wharves, estuaries, harbours.

Exclusions

- Work in offshore waters.

Major hazards

Refer to the Introduction for details of accident types and health risks

Immersion
Cold water shock.
Hypothermia.
Fatigue.
Drowning.

Moving water
River currents, scour.
Tidal movements.
Flood water.

Services
Electrocution/explosion,
(power cables in water).

Health hazards
Water-borne diseases
(eg Weil's Disease).
Contaminated water.
Chemicals.

Crushed/trapped
Impact by boats and pontoons.
Struck by plant or slewed materials and equipment.

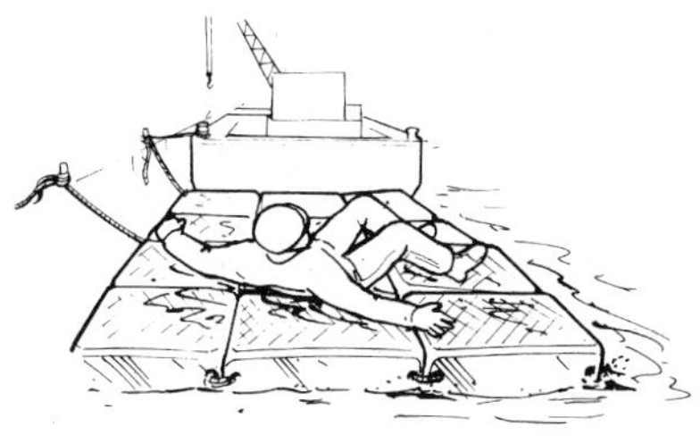

Falls
Slipping on wet surfaces.
Awkward access.
Unstable platforms.

Group E – Civil engineering
6 – Working over/near water

Specific hazard identification

Possible key considerations:

- **Site survey** - What are tidal movements? Any locks, weirs and sluices, etc?
- **Access/egress** - What is required to make safe with sufficient protection?
- **Working room** - What plant to use? Where to locate plant and materials?
- **Location** - What is the nature of the water (running or still)? Is it sea, river, canal, lake? What is depth and tidal range?
- **Working level** - What are the methods to safeguard falling from height? What emergency procedures are necessary?
- **Any others?**

Prompts

- Guard rails/barriers
- Protection of public
- Contamination
- Illumination
- Weather conditions
- Seasonal changes

Hazards consideration in design

Stage	Considerations and issues	Possible design options to avoid or mitigate hazards identified for working over/near water
Concept design	• Site survey • Water type/depth • Location • Services • Geotechnical investigation • Working space • Contamination • Safe access • Advance works (services, drainage) • Excavations • Emergency procedures • Sequence of operations	• Choose a prefabricated construction to avoid/minimise fabrication over water. • Divert or exclude the water by dams and pumping. • Consider temporary infilling or a grillage of pontoons to exclude or isolate the water. • Check whether working space or bed conditions exclude any construction methods.
Scheme design		• Ascertain a programme and design for the works to be carried out at low tide only. • Maximise the types of component which could be prefabricated. • Examine sequences of working to minimise period of exposure.
Detail design		• Select more durable materials and finishes for surfaces which require maintenance access over water (i.e. fewer visits required). • New works to include provision for future maintenance access where possible.

Examples of risk mitigation (methods of solution)

ACTION	ISSUE: Excavation in marshland
Avoidance Design to avoid identified hazards but beware of introducing others	No excavation would be required for a structure designed to be fabricated off-site and then ballasted to "float" at the required level.
Reduction Design to reduce identified hazards but beware of increasing others	The use of prefabricated tanks as the base units for all the large structures of the development would reduce excavation to the link buildings/corridors.
Control Design to provide acceptable safeguards for all remaining identified hazards	Design a cofferdam for the excavation of material from within. The internal water level would be controlled by pumping and the cofferdam could be incorporated into the permanent works.

Examples of risk mitigation (issues addressed at different stages)

A small road bridge is required across a shallow but navigable canal used by narrow boats. A lock is immediately downstream. The towpath is in constant use.

Concept design	Scheme design	Detail design
Review the options for a pre-fabricated/pre-cast bridge or components to minimise the work on site.	Match the size and weight of precast components to a practicable size and location of mobile crane.	Design piles which can be installed by piling plant without intruding into the towpath and footway access. Allow space for safe delivery and stockpile of materials.

A new suspension bridge is to cross a wide river. The steel prefabricated hollow section deck is to carry a 6 lane highway. Water clearance is 30m. A suitable road lighting system is to be provided.

Concept design	Scheme design	Detail design
Appropriate illumination could be provided from edge of deck or centrally. Coordinate spacing of lights with cables and hangers and consider access for future maintenance.	Consider "drop down" columns for easier maintenance, and a permanent barrier protected corridor which accesses every lamp post.	Make the whole lamp post, lamp unit, and wiring by off-site fabrication, so that only deck level connections are necessary on site.

Group E – Civil engineering
6 – Working over/near water

Related issues

References within this document

References	Related issues
A3	Access/egress routes for materials delivery and waste disposal will be determined by local conditions.
A4	Ground investigation will identify ground water levels and water quality.
A5	Site layout will be determined by project under construction in relation to space available (water authority may have specific site rules).
B1-B7	Excavation, drainage effects and support solutions could be crucial. Dewatering/exclusion of water will be important considerations.
C1-C8	Construction components and materials may require special routes within the site and storage may also be a problem.
D8	Electrical services mean electrocution is a major hazard when working near water.
E9	Related issues arise

References for further guidance

Primary general references and background information given in Section 4

Primary

- Construction Industry Training Board (CITB), Construction site safety notes No.30 GE 700/30 Working over water
- Building Employers Confederation (BEC), Construction safety, Section 8E Working over water 1988
- CIRIA SP137 (1997) Site safety for the water industry
- Morris, M and Simm, J (2000) Construction risk in river and estuary engineering, A guidance manual (Thomas Telford)

Secondary

- CIRIA SP57 (1998) Handling of materials on site

Background

- BS 8903:1991 Code of practice for the use of safety nets containment nets and sheets on constructional works

Classification

CI/SfB Code (--)13

Read the Introduction before using the following guidelines

Scope

- Cofferdams (in water, near water, in saturated ground).

Exclusions

- Cofferdams in off-shore environments (generally subject to extensive wave pressures and aggressive climate).
- Pressurised working.

Major hazards

Refer to the Introduction for details of accident types and health risks

Crushed/trapped
Collisions between boats, pontoons, crane barges, and cofferdam.
Moving plant and slewed equipment in confined space.

Falls
From dam wall into excavation or into water.

Inundation
Failure of cofferdam walls (due to water pressure/
scour/impact). Base failure (piping/heave).

Water
Cold water shock.
Tidal movements.
Currents and scour.
Drowning.

Health hazards
Contaminated ground water. Weil's Disease. Noxious gases in confined spaces. Noise and vibration from piling.

Services
Cables in river-bed. Temporary power cables in water. Fuel lines and storage.

Group E – Civil engineering
7 – Cofferdams

A GENERAL PLANNING
B EXCAVATIONS AND FOUNDATIONS
C PRIMARY STRUCTURE
D BUILDING ELEMENTS AND BUILDING SERVICES
E 7 CIVIL ENGINEERING

Specific hazard identification

Possible key considerations:

- **Project** - Can the cofferdam be wholly incorporated into the permanent structure?
- **Site investigation** - What is water table variation and extent of ground contamination? What is the underwater topography? Are there any obstructions?
- **Access** - What constraints affect movement of personnel, plant and machinery and materials delivery?
- **Adjacent water** - Is water running/still, navigable, liable to level variation, tidal, marine?
- **Environment** -What adjacent features/structures may be affected?
- **Working height** - Minimised by prefabrication of cofferdam?
- **Removal** - Will the permanent works obstruct the removal of temporary cofferdams?
- **Any others?**

Prompts

- Project programmes
- Ground conditions
- Water levels/depths
- Contaminants/health
- Illumination
- Weather conditions
- Emergency escape routes
- Safe access
- Services
- Removal/reinstatement
- Maintenance

Hazards consideration in design

Stage	Considerations and issues	Possible design options to avoid or mitigate hazards identified for cofferdams
Concept design	• Why the cofferdam? • Underwater survey • Adjacent water feature • Access • Environment • Ground investigation • Contamination • Programme of works • Working height • Support system • Stability • Services • Noise	• Impact by water traffic and floating debris may be resisted by the cofferdam or separate outer defences. • Consider the wider value of a temporary causeway instead of floating rigs. • Prefabricated caissons towed into place are quieter and involve less work in the water than driven piled cofferdams.
Scheme design		• Check the working platform/causeway level with respect to high tide/flood records and the intended season of works. • Provide scope for structural enhancement to respond to adverse monitoring reports. • Design plan shape of cofferdam to suit working space and construction plant as well as final structure.
Detail design		• Design out the need for underwater working by using a different construction or remotely controlled techniques. • Design in features for safe access, (eg holes and lugs for quick and easy fixing of equipment). • Coordinate the design of in-situ walings with permanent structure to minimise the extent of breaking out.

Examples of risk mitigation (methods of solution)

ACTION	ISSUE: Floating cranage and piling plant
Avoidance Design to avoid identified hazards but beware of introducing others	The use of heavy floating plant could be avoided by building the cofferdam off-site on land or in a dry dock. Instead, tug-boats would manoeuvre the floating cofferdam into position for sinking into place.
Reduction Design to reduce identified hazards but beware of increasing others	For a multi-span bridge over water, the sites of foundations in shallower waters could be accessed by building a temporary causeway for land-based plant to construct the cofferdams.
Control Design to provide acceptable safeguards for all remaining identified hazards	The project planning and contract periods are based on the assumption that the use of floating plant is only practicable at the turn of the tide and in good weather.

Examples of risk mitigation (issues addressed at different stages)

A large cofferdam is required for a bridge foundation in a river.

Concept design	Scheme design	Detail design
Consider permanent structure to be constructed. Consider effects on local marine movements to determine working space available.	Assume space available for double skin earth filled parallel walls. Provides safer working area, no internal strutting.	Wall design, drainage design, need for cross walls, include flexible and convenient details for the addition of working platforms.

A warehouse with deep basement storage is to be constructed adjacent to a navigable river - excavation for the foundation of the building is in saturated alluvial material.

Concept design	Scheme design	Detail design
Consider adjacent structures and environment to determine appropriate cofferdam type. Assume an external support.	Multi-level ground anchors or raking piles (instead of walings) provide safer working areas by being less intrusive into the confined space.	Consider the stability of adjacent structures and river bank, and review options for ground works in the intervening space to minimise their risk exposure. Include ground water and earth pressure monitoring devices to guide future maintenance/remedial work.

A GENERAL PLANNING
B EXCAVATIONS AND FOUNDATIONS
C PRIMARY STRUCTURE
D BUILDING ELEMENTS AND BUILDING SERVICES
E 7 CIVIL ENGINEERING

Related issues

References within this document

References	Related issues
A1,B6	Local environment is affected by piling which gives rise to noise and vibration. The effects on local buildings, industry and the general public must be considered.
A3	Access/egress routes for materials delivery and waste removal may also be dictated by local environment.
A4	Ground investigation is an important consideration in choice of support system.
B1,B2	Excavations inside cofferdams mean crucial considerations include ground conditions, drainage effects, support solutions and dewatering.
C1-C8	Foundations and Primary Structure may be constructed inside cofferdams.
D8	Electrical services mean electrocution is a major hazard when working near water.
E6	Working near water will involve additional hazards.

References for further guidance

Primary general references and background information given in Section 4

Primary

- CIRIA SP95 The design and construction of sheet piled cofferdams 1993

Secondary

- CIRIA SP57 (1988) Handling of materials on site
- Tomlinson, M J (2001) Foundation design and construction (Prentice Hall)
- Corus (2001) Piling Handbook

Background

Classification

CI/SfB Code (--)187

Group E – Civil engineering
8 – Pipes and cables

Read the Introduction before using the following guidelines

Scope

- Pipes: carrying water, sewage, gas, oil, etc. or used as protection for cables and "piped" services.
- Cables: for power, telephone, TV, other communications cables.

Exclusions

- Power cables for railways.
- Trenches (see B3).

Major hazards

Refer to the Introduction for details of accident types and health risks

Falls
From construction and maintenance plant and equipment.
Into trenches and excavations.

Buried/crushed/trapped
Struck by plant and machinery.
Collapse of trench or excavation.

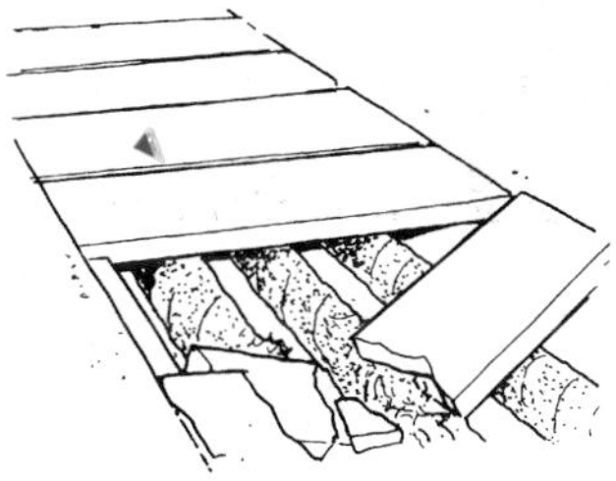

Services
Power – electrocution.
Gas and fuel – explosions.
Sewers – health hazards and sharps.

Health hazards
Contaminated ground and water (eg Weil's Disease).
Sewage and other effluents.
Noise.
Dust.

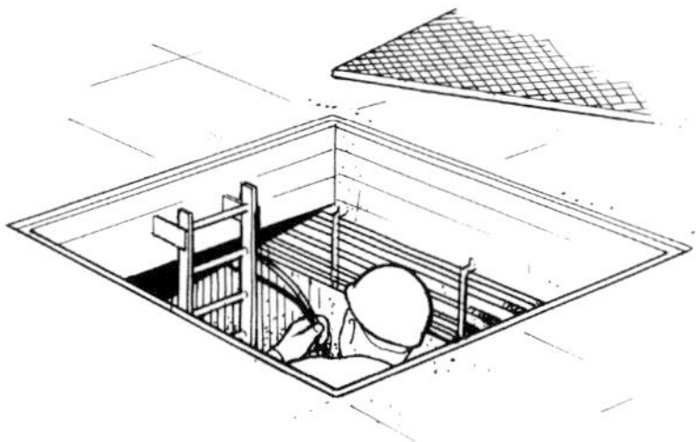

Asphyxiation
Working in confined spaces and access chambers.
Harmful gases.

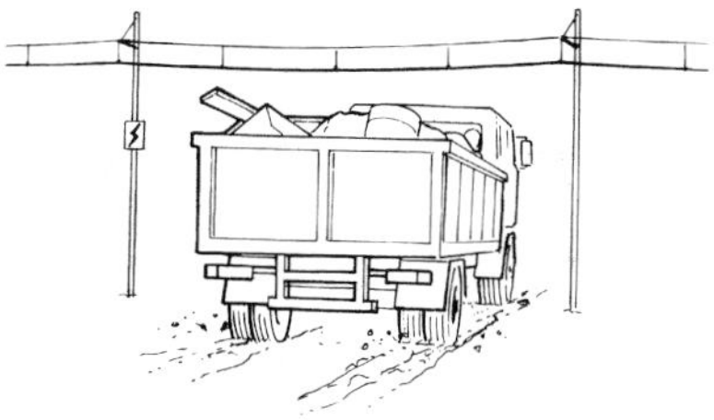

Traffic
Cabling over roads/
railways.
Runaway vehicles entering area of work.
Access chambers in the highway.

Specific hazard identification

Possible key considerations:

- **Facility** - What type of pipe or wire? What is it conveying?
- **Site survey** - What are site conditions? What other services are present (overhead or underground)? How high, how deep are they? What ground contaminants are present?
- **Working room** - Is there sufficient space for plant and machinery?
- **Environment** - Is work next to structures, buildings, road, railway, also near to water? Special procedures? Need for double protection for containment?
- **Access:** What site access route problems exist? Is there room to work on new systems installation?
- **Interface** - With utilities and services to buildings/structures.
- **Any others?**

Prompts

- Ground conditions
- What service?
- Contamination
- Access
- Possession
- Maintenance
- Confined spaces (access chambers)
- Party wall interfaces
- Existing systems
- Trace heating
- Lagging
- Access for maintenance
- Provision for future adaptation

Hazards consideration in design

<table>
<tr><th>Stage</th><th>Considerations and issues</th><th>Possible design options to avoid or mitigate hazards identified for pipes and cables</th></tr>
<tr><td>Concept design</td><td rowspan="3">• Site location
• Existing information
• Access
• Possession
• Type of pipe/cable
• Height and depth
• Services
• Ground conditions
• Adjacent features
• Sequence
• Working space
• Maintenance
• Local environment (type of conduit/ temperature control)
• Other services
• Cleaning and access requirements
• Build-up of gases
• Access for CCTV
• Exclusion of public
• Permanent access</td><td>• Review current data to determine the benefit of tracing or surveying some of the existing services.
• Review the possibility of co-ordinating the work with planned work by other utilities to streamline construction operations and reduce public disruption.
• Investigate existing services fully.
• Concept design to minimise need for access into chambers etc for cleaning.</td></tr>
<tr><td>Scheme design</td><td>• Plan the route to avoid existing services and minimise the likelihood of disturbance by other construction activity.
• Ensure that the design provides for easy location and access to make future connections.
• Provide ventilation of enclosed spaces.
• Arrange layout to assist access to valves.
• Provide permanent access and use during construction.</td></tr>
<tr><td>Detail design</td><td>• Whenever possible choose details which anticipate technical change.
• Balance the provision of convenient, safe access for future maintenance against easy access by untrained or unauthorised persons.
• Provide double-valve isolation for man access.
• Use V-J-type couplings to assist future work.
• Provide overhead lifting systems for heavy pipes and equipment.</td></tr>
</table>

Examples of risk mitigation (methods of solution)

ACTION / ISSUE	In-trench jointing of pipes	Crossing a major road
Avoidance Design to avoid identified hazards but beware of introducing others	Select a pipe/duct capable of being laid from a "reel" located above ground.	Route the service over an existing bridge or micro-bore under the road.
Reduction Design to reduce identified hazards but beware of increasing others	Choose quick and simple jointing (eg push-fit or hinged clamps instead of lead filled or welded connections).	Connect excavations on each side of the carriageway by pipe-jacking a service duct between them.
Control Design to provide acceptable safeguards for all remaining identified hazards	Locate complicated joints where wider excavation is possible and likely.	Design for construction in short lengths to suit an acceptable sequence of lane closures.

Examples of risk mitigation (issues addressed at different stages)

A new sewer is to be laid in an urban environment. Records have identified underground power cables which cross the alignment and sometimes run parallel to the alignment.

Concept design	Scheme design	Detail design
Re-examine route of sewer to avoid underground services.	Prove information by trial pits and electro-magnetic tracing. Determine the width of working corridor required. Consider use of trenchless techniques. Route cannot be changed.	Trenchless techniques chosen. Specify location of access pits and alignment of sewer to avoid existing services.

To release development land an underground network of pipes for power station cooling water has to be partially replaced and diverted.

Concept design	Scheme design	Detail design
Consider the future maintenance benefit of a whole new network. Review alternative routes and combinations of overhead and underground alignments.	Favour options which allow complete construction of the new routes before breaking into the existing network.	Maintain access to live services during construction. Define the working zone to minimise overlap between construction activities and the operation of the power station.

Group E – Civil engineering
8 – Pipes and cables

GENERAL PLANNING A
EXCAVATIONS AND FOUNDATIONS B
PRIMARY STRUCTURE C
BUILDING ELEMENTS AND BUILDING SERVICES D
CIVIL ENGINEERING E8

Related issues

References within this document

References	Related issues
A1	Surrounding environment will be affected by pipe and cable work (including maintenance operations).
A2	Demolition of a structure must include identification and marking of services that are isolated or redundant. The Health and safety file should be available.
A4	Access/egress routes and facilities for maintenance should be free from obstruction.
B3	Trench excavations must be suitably supported.
D8	Electrical services.
E2	Working space for pipes and cables are invariably adjacent to or under roadways. When traffic measures are required during construction or maintenance, consider space and barriers to protect site personnel.

References for further guidance

Primary general references and background information given in Section 4

Primary

- HSE HSG47 Avoiding danger from underground services
- HSE GS6 Avoidance of danger from overhead electric power lines
- HSE CIS8 (revised) Safety in excavations
- CIRIA SP57 Handling of materials on site 1988

Secondary

- CIRIA R97 Trenching practice 1992
- CIRIA TN127 Trenchless construction for underground services 1987
- HSE MDHS100 Surveying sampling and assessment of asbestos coating materials

Background

- DoT Chapter 8 Traffic Signs Manual 2002
- HSE INDG84L – Leptospirosis: are you at risk?
- Pipeline Industries Guild (1994) Pipelines: all you wanted to know… but were afraid to ask (Pipeline Industries Guild)
- Hughes, H (1997) Telecommunications cables: manufacture and installation (John Wiley)

Classification

CI/SfB Code (--)I/J

Group E – Civil engineering
9 – Work in coastal and maritime waters

Read the Introduction before using the following guidelines

Scope

- Exposed Tidal waters (eg Coastal waters, harbours, estuaries) where wave, wind and currents impact on the construction methodology, plant band environment.

Exclusions

- Working over/near water (See E6)

Major hazards

Refer to the Introduction for details of accident types and health risks

Slips, trips and falls due to:
Wet, algae-covered surfaces.
Awkward access.
Unstable and moving platforms.
Moving water.
Exposed locations.

Crushed/trapped due to:
Moving plant.
Inter-vessel personnel transfer.
Impact by boats and pontoons.
Struck by plant or slewed materials and equipment.

Health hazards due to:
Lack of accessible welfare provisions due to exposed/remote location
Contaminated water:
- Water-borne diseases (eg Weill's Disease).
- Chemicals contamination

Immersion
Cold water shock
Hypothermia
Fatigue
Drowning

Moving water
River currents, scour
Tidal movements.
Flood water

Unsafe temporary structures
Unconsolidated seabed material.
Seabed changes
Temporary storage eg rock dump

Group E – Civil engineering
9 – Work in coastal and maritime waters

Specific hazard identification

Possible key considerations:

- **Don't ignore the normal hazards** (eg slips trips and falls) but consider how the environment might affect them
- **Site data/location:**
  - What are tidal and surge conditions?
  - What are the wave conditions?
  - What are the flow conditions?
  - What are the possible changes in ground/seabed level and condition?
- **Access/egress** – What is required/practicable to make safe? What are the methods for transferring personnel to vessels?
- **Working room** – What plant will be used? Where will plant and materials be located?
- **Working level** – What are the methods to safeguard falling from height? What measures are in place to recover personnel (including suspension after falls)?
- **The public** – how will the public be excluded from working/unsafe areas?
- **Any others?**

Prompts

- Guard rails/barriers
- Protection of public
- Contamination
- Illumination
- Weather conditions
- Seasonal changes
- Ground conditions
- Welfare facilities

Hazards consideration in design

Stage	Considerations and issues	Possible design options to avoid or mitigate hazards identified for deep basements and shafts
Concept design	• Site survey • Water type/depth • Location • Services • Geotechnical investigation • Working space • Contamination • Safe access • Advance works (services, drainage) • Excavations • Emergency procedures • Sequence of operations	• Choose a prefabricated construction to avoid/minimise fabrication over water. • Divert or exclude the water by dams and pumping. • Consider cofferdams to exclude or isolate the water hazards. • Check whether working space or bed conditions exclude any construction methods.
Scheme design		• Ascertain a programme and design for the works to be carried out at low tide only. • Maximise the types of component which could be prefabricated. • Examine sequences of working to minimise period of exposure.
Detail design		• Select more durable materials and finishes for surfaces which require maintenance access over water (i.e. fewer visits required). • New works to include provision for future maintenance: safe removable access where possible. • Remove/reduce underwater fixings to avoid diver-placement

Examples of risk mitigation (methods of solution)

ACTION \ ISSUE	Constructing under water formations
Avoidance Design to avoid identified hazards but beware of introducing others	No excavation would be required for a structure designed to be fabricated off-site and then ballasted to "float" at the required level.
Reduction Design to reduce identified hazards but beware of increasing others	The use of prefabricated tanks as the base units for all the large structures of the development would reduce excavation to the link buildings/corridors
Control Design to provide acceptable safeguards for all remaining identified hazards	Design a cofferdam for the excavation of material from within. The internal water level would be controlled by pumping and the cofferdam could be incorporated into the permanent works.

Examples of risk mitigation (issues addressed at different stages)

An old Victorian pier is being refurbished in a busy seaside resort.

Concept design	Scheme design	Detail design
Review the options for a pre-fabricated/pre-cast solution or components to minimise the work on site.	Minimise the number of precast components whilst using a practicable crane size. Allow space for safe delivery and stockpile of materials. Remove underwater fixings where possible to avoid diver working.	Allow removal of existing fittings for repair in workshop environment rather than repair over water. Provide access/fixing points for ease of maintenance/painting.

A costal protection scheme is required to protect an urban area involving beach recharge and groynes

Concept design	Scheme design	Detail design
Risk assessment to compare: • Placement of sand recharge • Repairing groins in tidal areas • Offshore breakwater	Required placing methodology to avoid soft ground areas which may be hazardous to public.	Consider using standard timber details rather than varying details to limit amount of cutting work required.

Group E – Civil engineering
9 – Work in coastal and maritime waters

A GENERAL PLANNING
B EXCAVATIONS AND FOUNDATIONS
C PRIMARY STRUCTURE
D BUILDING ELEMENTS AND BUILDING SERVICES
E 9 CIVIL ENGINEERING

Related issues

References within this document

References	Related issues
A4	Ground investigation will identify ground water levels and water quality.
A3	Access/egress routes for materials delivery and waste disposal will be determined by local conditions.
A5	Site layout will be determined by project under construction in relation to space available (water authority may have specific site rules).
B1–B7	Excavation, drainage effects and support solutions could be crucial. Dewatering/exclusion of water will be important considerations.
CI–C8	Construction components and materials may require special routes within the site and storage may also be a problem.
D8	Electrical services mean electrocution is a major hazard when working near water.
E6	Related issues arise.

References for further guidance

Primary general references and background information given in Section 4

Primary

- Construction Industry Training Board, Construction site safety notes No.30 GE 700//30 Working over water (1992)
- Building Employers Confederation, Construction safety, Section 8E Working over water (1988)
- Morris, M and Simm, J (2000) Construction risk in river and estuary engineering, A guidance manual (Thomas Telford)
- CIRIA C518 Safety in ports, ship-to-shore linkspans and walkways
- Moth, P (1998) Work boat code of practice – an operational guide (Forecourt Publications)

Secondary

- CIRIA SP130 (1997) Site safety for the water industry
- CIRIA R158 (1996) Sea outfalls – inspection and diver safety
- CIRIA R159 (1996) Sea outfalls – construction inspection and repair. An engineering guide
- HSE HSG177 Managing health and safety in dockwork

Background

- CIRIA C584 Coastal and marine environmental site guide (2003)
- Cork R S, Cruickshank I C (In prep) Construction safety in coastal and maritime engineering (Thomas Telford)

Classification

CAWS Group A, CEWS Class T,

3 Documenting design decisions on individual projects

3.1 THE PROVISION AND USE OF HEALTH AND SAFETY INFORMATION

This section provides additional guidance in support of the information in Section 1.4.13.

Designers have a duty to provide information on the significant risks in the design that cannot be avoided (see 3.2 below). In addition, designers may wish to document the health and safety considerations made during the design, to keep track of hazards that have not yet been resolved, to indicate the range of design options that were considered at any stage (see 3.3 below), or to communicate information to the planning superviser and other designers.

Documents should not be allowed to create a significant burden of extra paperwork; the ACoP (para 6) emphasizes that paperwork must not be for its own sake and that bureaucracy must be minimized. Paperwork should be used as an aid to thinking, communication and recording what has been done and as an aid to any audits or investigations that may occur. Many organizations already use management systems such as quality assurance, and undertake design reviews; recording the consideration of health and safety during design will form one part of any such system.

3.2 STATUTORY INFORMATION FOR REGULATION 13(2)(b)

The provision by designers of drawings and specifications is normal. Additionally, other statements or special reports or both may be needed when:

- the work to be done poses potential risks to people in adjacent structures
- the work to be done poses potential risks to the public or the client's own employees or customers
- the structure is of a nature that might pose health and safety risks to competent contractors (for example, having fragile roof materials)
- designers have made assumptions about the likely method of work and sequence of construction that contractors will need to know in order to assess what needs to be done and to devise safe working methods; the provision of a method statement may be a useful way to communicate the hazards to contractors.

A designer will need to understand how each structure can be constructed or assembled safely. When a designer does not have the required specialist knowledge, it will be necessary to consult with the planning supervisor or with specialists. Recognition of one's own limits and access to advice is part of a designer's competence.

Information about the design will be communicated in various forms including:

- drawings
- outline method statements
- specifications/scope of works
- special report(s).

These can be included as tender documentation. However, it should normally be made clear that the information does not preclude a contractor's consideration of other options. Designers should provide relevant information and will wish to co-operate with contractors. Contractors may consult designers, but contractors will be responsible for construction methods they develop as alternatives to suggestions from the designers.

Exceptions to the above might include circumstances where it is intended to impose a particular method of construction on a contractor, eg when designers have assumed a particular sequence of erection to prevent buckling of the partially complete structure or when a safe method of working has already been approved by a responsible body. In such cases tender documents should require contractors to confirm that the proposed method(s) are acceptable and they are able to manage them safely on site. These issues should be clear within the pre-tender health and safety plan. The competence and resources to manage the residual risk will be key factors influencing the selection of any contractor involved in this aspect of the project.

3.3 INFORMATION FOR IN-HOUSE RECORDS

Designers may like to keep their own records of the hazards they have identified and how they have dealt with the associated risks. These will be discussed during design reviews and progress meetings and the appropriate records could form part of the existing systems for project management. There will be practical reasons to keep records, in case of personnel changes, unexpected delays in the progress of the design, or in the case of future dispute.

This guidance does not prescribe how designers should record their considerations and decisions arising from compliance with the CDM Regulation, as each practice will have its own approach when establishing procedures and audit trails.

Designers may find it helpful to develop standard forms for recording the health and safety decisions for their projects. This may include the actions they have been able to take to avoid hazards, reduce levels of risk or provide control measures. Such forms may also include a consideration by the designer of whether the residual risk is sufficiently significant or unusual to be outside the scope of a competent contractor in the context of the project, and therefore a candidate for inclusion in the health and safety plan or file.

4 Glossary

This section gives explanations of the common terms used in this document. Where explanations have been extracted from existing documents this is noted.

AGENT	A person whose trade, business or other undertaking (whether for profit or not) is to act as an agent for a client. Employees of the client who discharge functions on behalf of the client are not agents of the client for the purposes of this definition.
ACoP	See *Approved Code of Practice*.
APPROVED CODE OF PRACTICE (ACoP)	Health and safety codes of practice are approved by the Health and Safety Commission and have special legal status. The code associated with the CDM Regulations is contained within "Managing Health and Safety in Construction" : Construction (Design and Management) Regulations 1994 (ref.HSG224, by HSE Books, referred to as "the ACoP"). See also introductory text on page 3 of this guide.
CLEANING WORK	Cleaning work is a term defined in the Regulations as follows: "The cleaning of any window or any transparent or translucent wall, ceiling or roof in or on a structure, where such cleaning involves a risk of a person falling more than 2 metres." [*Reg 2(1)*] However, the health and safety implications of other types of cleaning work must also be considered.
CLIENT	"Any person for whom a project is carried out, whether it is carried out by another person (a client's agent) or is carried out in-house." [*Reg 2(1)*]
CLIENT'S AGENT	A "client's agent" for the purposes of CDM is any agent or other client appointed to act as the only client. The appointment may be "declared" to the HSE, or not. [*Based on Reg 4*]
CONSTRUCTION PHASE	"The period of time commencing when construction work in any project starts and ending when construction work in that project is completed." [*Reg 2(1)*]
CONSTRUCTION RISK ASSESSMENT	The process of risk assessment (see "risk assessment" below) applied by contractors to the construction process.
CONSTRUCTION WORK	"The carrying out of any building, civil engineering or engineering construction work and includes any of the following: • the construction, alteration, conversion, fitting out, commissioning, renovation, repair, upkeep, redecoration or other maintenance (including cleaning which involves the use of water or an abrasive at high pressure or the use of substances classified as corrosive or toxic for the purposes of

Regulation 7 of the Chemicals (Hazard Information and Packaging) Regulations 1993), decommissioning, demolition or dismantling of a structure

- the preparation for an intended structure, including site clearance, exploration, investigation (but not site survey) and excavation, and laying or installing the foundations of the structure
- the assembly of prefabricated elements to form a structure or the disassembly of prefabricated elements which, immediately before such disassembly, formed a structure
- the removal of a structure or part of a structure or of any product or waste resulting from demolition or dismantling of a structure or from disassembly or prefabricated elements which, immediately before such disassembly, formed a structure, and
- the installation, commissioning, maintenance, repair or removal of mechanical, electrical, gas, compressed air, hydraulic, telecommunications, computer or similar services, which are normally fixed within or to a structure.

but does not include the exploration for or extraction of mineral resources or activities preparatory thereto carried out at a place where such exploration or extraction is carried out." [*Reg 2(1)*]

CONTRACTOR

"Any person who carries on a trade, business or other undertaking (whether for profit or not) in connection with which he:

(a) undertakes to or does carry out or manage construction work

(b) arranges for any person at work under his control (including, where he is an employer, any employee of his) to carry out or manage construction work." [*Reg 2(1)*]

This includes subcontractors, main contractors, trade contractors, turnkey contractors and design-and-build contractors. It may also be interpreted as including others who manage (but do not contract to carry out) construction work.

COMPETENCE

Having, or having ready access to the skills, knowledge, experience, systems and support necessary to carry out work relating to the construction work in hand, in a manner that takes due account of health and safety issues.

CONTROL

Measure taken to reduce (or mitigate) risk.

DBFO

Design, build, finance and operate; a form of procurement.

DECLARED CLIENT'S AGENT

See *client's agent*.

DEMOLITION AND DISMANTLING	Demolition and dismantling is defined in the ACoP (para 30) as including the deliberate pulling down, destruction or taking apart of a structure (and) dismantling for re-erection or re-use.
DESIGN	"Design in relation to any structure includes drawings, design details, specifications and bill of quantities (including specification of articles or substances) in relation to the structure." [*Reg 2(1)*]
DESIGNER	"Any person who carries on a trade, business or other undertaking in connection with which he prepares a design relating to a structure or part of a structure." [*Reg 2(1)*] This will include not only design professionals but also others who make decisions about materials or how construction work will be done.
DESIGN RISK ASSESSMENT	The process of risk assessment (see *risk assessment* below) applied by designers to their design.
DEVELOPER	For the purposes of the CDM Regulations, a developer has a special meaning (see Reg 5), "when commercial developers sell domestic premises before the project is complete and arrange for construction work to be carried out", whereby the Regulations apply as if the developer were client and the work is not treated as being for a domestic client. In other words, a commercial housing developer is the client for the housing development even though he may have sold some or all of the properties before completion of the works. See also ACoP paras 101, 102.
DOMESTIC CLIENT	"A client for whom a project is carried out not being a project carried out in connection with the carrying on by the client of a trade, business or other undertaking (whether for profit or not)." [*Reg 2(1)*] See also ACoP paras 94, 95.
FIXED PLANT	Fixed plant means (for the purposes of CDM) plant and machinery that is used for a process; it is not part of the services associated with the structure or building services.
HAZARD	Something with the potential to cause harm
HEALTH AND SAFETY FILE	This is a record of information for the client (and others who need to see it), which focuses on health and safety. It alerts those who are responsible for the structure and equipment in it of the significant health and safety risks that will need to be dealt with during subsequent use, construction, maintenance, repair and cleaning work.

HEALTH AND SAFETY PLAN	This is a document that contains information to assist with the management of health and safety as the project proceeds. It has two main stages, pre-tender and construction. The pre-tender health and safety plan (so named because it is normally prepared before the tendering process for the construction contract) brings together the health and safety information obtained from the client and designers. The construction health and safety plan details how the construction work will be managed on site to ensure health and safety.
HSC	Health and Safety Commission.
HSE	Health and Safety Executive.
MAINTENANCE	Construction work done to keep a structure functioning (see also ACoP glossary).
METHOD STATEMENT	A written document laying out work procedures and sequence of operation. It takes account of the risk assessment carried out for the task or operation and the control measures identified. The term "safety method statement" has been used but is not a term normally used in the construction industry. All method statements should take account of safety issues.
NOTIFIABLE	Construction work is notifiable to the HSE if CDM applies and if the construction phase (including commissioning) is expected to last more than 30 working days or will involve more than 500 person days of work.
PERSON	A "person" may be an individual, a company or a corporate entity.
PFI	Private finance initiative, a form of public-sector procurement.
PPP	Public-private partnership, a form of public-sector procurement.
PLANNING SUPERVISOR	The Regulations define the planning supervisor as a person who carries out the function as defined. The planning supervisor is the person who co-ordinates and manages the health and safety aspects of design. The planning supervisor also has to ensure that the pre-tender stage of the health and safety plan and the health and safety file are prepared.
PRINCIPAL CONTRACTOR	A role prescribed by the Regulations. The principal contractor is a contractor who is appointed by the client. The principal contractor has the overall responsibility for the management and co-ordination of site operations with respect to health and safety.
PROJECT	This means "a project that includes, or is intended to include, construction work." [*Reg 2 (1)*]
PROJECT MANAGER	For the purpose of this guide, the project manager is the party responsible for the management of the project on behalf of the client. He may also be referred to by other terms such as "contract administrator" or "client's representative".

RISK	The likelihood that harm from a particular hazard will occur and the possible severity of the harm.
RISK ASSESSMENT	The process of identifying hazards, assessing the degree of risk associated with them and identifying suitable control measures (see *control* above).
SAFETY METHOD STATEMENT	See *method statement* above.
SO FAR AS REASONABLY PRACTICABLE	To carry out a duty "so far as reasonably practicable" means that the degree of risk in a particular activity can be balanced against the time, trouble, cost and physical difficulty of taking measures to avoid the risk. If these are so disproportionate to the risk that it would be quite unreasonable for the people concerned to have to incur them to prevent it, they are not obliged to do so. The greater the risk, the more likely it is that it is reasonable to go to very substantial expense, trouble and invention to reduce it. However, if the consequences and the extent of a risk are small, insistence on great expense would not be considered reasonable.
SPV	Special purpose vehicle, a company set up to carry out a PFI project.
STRUCTURE	This means: • any building, steel or reinforced concrete structure (not being a building), railway line or siding, tramway line, dock, harbour, inland navigation, tunnel, shaft, bridge, viaduct, waterworks, reservoir, pipe or pipe-line (whatever, in either case, it contains or is intended to contain), cable, aqueduct, sewer, sewage works, gasholder, road, airfield, sea defence works, river works, drainage works, earthworks, lagoon, dam, wall caisson, mast, tower, pylon, underground tank, earth retaining structure, or structure designed to preserve or alter any natural feature, and any other structural similar to the foregoing; or • any formwork, falsework, scaffold or other structure designed or used to provide support or means of access during construction work; or • any fixed plant in respect of work that is installation, commissioning, decommissioning or dismantling and where any such work involves a risk of a person falling more than 2 metres. [*Reg 2(1)*]

5 Sources of futher information

The references given below and in each of the work sectors in Section 2 were correct at the date of publication. However, reference documents are liable to be updated, superseded or withdrawn and readers are urged to check that they have obtained the most up-to-date references. See also further references in the ACoP.

Publications

CONSTRUCTION INDUSTRY TRAINING BOARD (2003)
Construction Site Safety - Health, Safety and Environmental Information
GE700, CITB, King's Lynn

DAVIES, V J AND TOMASIN, K (1996)
Construction Safety Handbook
Thomas Telford, London

HEALTH AND SAFETY COMMISSION (2001)
Managing Health and Safety in Construction: Construction (Design and Management) Regulations 1994. Approved Code of Practice and Guidance
HSG224, HSE Books, Sudbury

HEALTH AND SAFETY EXECUTIVE (2000)
Management of health and safety at work. Management of Health and Safety at Work Regulations 1999. Approved code of Practice and Guidance (Second edition)
L21, HSE Books, Sudbury

HEALTH AND SAFETY EXECUTIVE (1996)
Health and safety in construction
HSG150, HSE Books, Sudbury

HEALTH AND SAFETY EXECUTIVE (1995)
Construction (Design and Management) Regulations 1994
- *The role of the client*, CIS39 (rev 2000)
- *The role of the planning supervisor*, CIS40 (rev 2000)
- *The role of the designer*, CIS41
- *The pre-tender health and safety plan*, CIS42
- *The health and safety plan during the construction phase*, CIS43
- *The health and safety file*, CIS44

HSE Books, Sudbury

FERGUSON, I (1995)
Dust and noise in the construction process
CRR73, HSE Books, Sudbury

SEDDON, P (2002)
Harness suspension: review and evaluation of existing information
CRR451, HSE Books, Sudbury

THE CONSTRUCTION CONFEDERATION AND CITB
Construction Health and Safety Manual (updated twice yearly)
Construction Industry Publications Ltd, Birmingham

CIRIA publications

ARMSTRONG, J (2002)
Facilities management manuals – a best practice guide,
C581, CIRIA, London

BIELBY, S C AND READ, J A (2001)
Site safety: a handbook for young construction professionals. 3rd edition
SP151, CIRIA, London

CIRIA (1993)
Not just an accident (Video),
SP100, CIRIA, London

CIRIA (1993)
A guide to the control of substances hazardous to health in design and construction,
R125, CIRIA, London

CIRIA (1999)
CDM Training Pack for Designers,
C501, CIRIA, London

CIRIA (1999)
Operating and maintenance manuals for buildings, a guide to procurement and preparation,
C507, CIRIA, London

CIRIA (2000)
Integrating safety, quality and environmental management,
C509, CIRIA, London

DELVES, A; DRAYTON, R AND SHEEHAN, T (2002)
Faster construction on site by selection of methods and materials,
C560, CIRIA, London

EVANS, D; JEFFERIS, S A; THOMAS, A O AND CUI, S (2001)
Remedial processes for contaminated land – principles and practices,
C549, CIRIA, London

IDDON, J AND CARPENTER, J (2003)
Safe access for maintenance and repair
C611, CIRIA, London

ILLINGWORTH, J AND THAIN, K (1988)
Handling of materials on site,
SP57, CIRIA, London

LLOYD, D (ed) (2003)
Crane stability on site: an introductory guide (second edition)
C703, CIRIA, London

LLOYD, D AND KAY, T (1995)
Temporary access to the workface: a handbook for young professionals,
SP121, CIRIA, London

RUDLAND, D J; LANCEFIELD, R M AND MAYELL, P A (2001)
Contaminated land risk assessment – a guide to good practice,
C552, CIRIA, London

WS ATKINS AND GILBERTSON, A (2004)
CDM Regulations – Practical guidance for clients and clients' agents,
C602, CIRIA, London

WS ATKINS AND GILBERTSON, A (2004)
CDM Regulations – Practical guidance for planning supervisors,
C603, CIRIA, London

British Standards

BS 8093:1991
Code of Practice for the use of safety nets, containment nets and sheets on construction work

Websites

<www.hse.gov.uk>
About the HSE

<www.hse.gov.uk/construction/designers>
HSE website devoted to construction designer issues

<www.hsebooks.co.uk>
Search for HSE publications

<www.safetyindesign.org.uk>
Safety in Design (SiD) website; includes:

- CDM guidance for designers (by Construction Industry Council (CIC))
- Accreditation of health and safety training for designers

<www.cbpp.org.uk>
Best practice site: search on health and safety

<wwt.uk.com>
Improved health and safety awareness campaign

<www.learning-hse.com>
HSE e-learning site; gives further links

<www.design4health.com>
Interactive guidance toolkit for designers

<www.aps.org.uk>
Association of planning supervisors

Appendix A
Checklist of hazard management

This checklist contains generic information. It is indicative, not exhaustive and must only be used to facilitate careful consideration of the specific hazards discussed and levels of risk on a specific project; see also Chapter 2 for guidance.

Hazard management **Issue**	**Response**
a) Select the position and design of structures to minimise risks from site hazards, including: • buried services, including gas pipelines • overhead cables • traffic movements to, from and around the site • contaminated ground, for example minimising disturbance by using shallow excavations, and driven, rather than bored, piles.	
b) Design out health hazards, for example: • specify less hazardous materials (eg solvent-free or low-solvent adhesives and water-based paints) • avoid processes that create hazardous fumes, vapours, dust, noise or vibration, including disturbance of existing asbestos, cutting chases in brickwork and concrete, breaking down cast *in-situ* piles to level, scabbling concrete, hand digging tunnels, flame cutting or sanding areas coated with lead paint or cadmium • specify materials that are easy to handle (eg lighter weight building blocks) • design block paved areas to enable mechanical handling and laying of blocks.	
c) Design out safety hazards, for example: • the need for work at height, particularly where it would involve work from ladders, or where safe means of access and a safe place of work is not provided • fragile roofing materials • deep or long excavations in public areas or on highways • materials that could create a significant fire risk during construction.	
d) Consider prefabrication to minimise hazardous work or to allow it to be carried out in more controlled conditions off-site including for example: • design elements, such as structural steel work and process plant, so that sub-assemblies can be erected at ground level and then safely lifted into place • arrange for cutting to size to be done off-site, under controlled conditions, to reduce the amount of dust release.	
e) Design in features that reduce the risk of falling/injury where it is not possible to avoid work at height, for example: • early installation of permanent access, such as stairs, to reduce the use of ladders • edge protection or other features that increase the safety of access and construction.	

Hazard management Issue	Response
f) Design to simplify safe construction, for example: • provide lifting points and mark the weight, and centre of gravity of heavy or awkward items requiring slinging both on drawings and on the items themselves • make allowance for temporary works required during construction • design joints in vertical structural steel members so that bolting up can easily be done by someone standing on a permanent floor, and by use of seating angles to provide support while the bolts are put in place • design connections to minimise the risk of incorrect assembly.	
g) Design to simplify future maintenance and cleaning work, for example: • make provision for safe permanent access • specify windows that can be cleaned from the inside • design plant rooms to allow safe access to plant and for its removal and replacement • design safe access for roof-mounted plant, and roof maintenance • make provision for safe temporary access to allow for painting and maintenance of facades etc. This might involve allowing for access by mobile elevating work platforms or for erection of scaffolding.	
h) Identify demolition hazards for inclusion in the health and safety file, for example: • sources of substantial stored energy, including pre- or post-tensioned members • unusual stability concepts • alterations that have changed the structure.	
i) Understand how the structure can be constructed, cleaned and maintained safely: • taking full account of the risks that can arise during the proposed construction processes, giving particular attention to new or unfamiliar processes, and to those that may place large numbers of people at risk • considering the stability of partially erected structures and, where necessary, providing information to show how temporary stability could be achieved during construction • considering the effect of proposed work on the integrity of existing structures, particularly during refurbishment • ensuring that the overall design takes full account of any temporary works, for example falsework, which may be needed, no matter who is to develop those works • ensuring that there are suitable arrangements (for example access and hard standing) for cranes, and other heavy equipment, if required.	
j) ANYTHING ELSE?	

Appendix B
Checklist of information to be provided

This checklist contains generic information. It is indicative, not exhaustive and must only be used to facilitate careful consideration of the specific information to be provided on a specific project; see also Section 1.4.13 and Chapter 3 for guidance.

CDM – Information **Issue**	**Response**
a) Designers must include adequate health and safety information with the design. This includes information about hazards that remain in the design, and the resulting risks. They need to make clear to planning supervisors, or whoever is preparing the pre-tender plan, any assumptions about working methods or precautions, so that the people carrying out the construction work can take them into account.	
b) Designers do not need to mention every hazard or assumption, as this can obscure the significant issues, but they do need to point out significant hazards. These are not necessarily those that result in the greatest risks, but those that are: • not likely to be obvious to a competent contractor or other designers • unusual • likely to be difficult to manage effectively. To identify significant hazards designers must understand how the design can be built.	
c) Examples of significant hazards where designers always need to provide information include: • hazards that could cause multiple fatalities to the public, such as tunnelling, or the use of a crane close to a busy public place, major road or railway • temporary works, required to ensure stability during the construction, alteration or demolition of the whole or any part of the structure, eg bracing during construction of steel or concrete frame buildings • hazardous or flammable substances specified in the design, eg epoxy grouts, fungicidal paints, or those containing isocyanates • features of the design and sequences of assembly or disassembly that are crucial to safe working • specific problems and possible solutions for example arrangements to enable the removal of a large item of plant from the basement of a building • structures that create particular access problems, such as domed glass structures • heavy or awkward prefabricated elements likely to create risks in handling • areas needing access where normal methods of tying scaffolds may not be feasible, such as facades that have no opening windows and cannot be drilled.	
d) Information should be clear, precise, and in a form suitable for the users. This can be achieved using, for example: • notes on drawings – these are immediately available to those carrying out the work, they can refer to other documents if more detail is needed, and be annotated to keep them up to date • a register, or lists of significant hazards, with suggested control measures • suggested construction sequence showing how the design could be erected safely, where this is not obvious, for example suggested sequences for putting up stressed skin roofs. Contractors may then adopt this method or develop their own approach.	
e) ANYTHING ELSE?	